Immersive Sound Production

T0266839

Immersive Sound Production is a handbook for the successful implementation of immersive sound for live sports and entertainment.

This book presents thorough explanations of production practices and possibilities and takes the reader through the essentials of immersive sound capture and creation with real world examples of microphones, mixing and mastering practices. Additionally, this book examines the technology that makes immersive sound possible for the audio mixer, sound designer and content producer to craft a compelling soundscape.

This book serves as a guide for all audio professionals, from aspiring audio mixers to sound designers and content producers, as well as students in the areas of sound engineering, TV and broadcast and film.

Dennis Baxter's over 37 years of progressive sound design includes 11 Olympic Games and hundreds of sporting events and has earned him a reputation as a globally recognized sound designer, netting him six Emmy® Awards along the way. He was an early adopter of surround sound for live events and ushered in 5.1 surround at every sport and venue at the 2008 Beijing Olympic Games. He has spent almost a decade preparing for immersive sound production.

Dennis is the author of *A Practical Guide to Television Sound Engineering* (Focal Press 2007), published in both English and Chinese. He also authors columns and articles for professional publications, and is a regular speaker to professional television industry groups. As an author and an educator, his mission has been to define and teach the core competencies necessary to be successful in broadcasting.

Immersive Sound Production

A Practical Guide

Dennis Baxter

Routledge
Taylor & Francis Group

LONDON AND NEW YORK

Cover image: filo / Getty Images

First published 2022
by Routledge
4 Park Square, Milton Park, Abingdon, Oxon OX14 4RN

and by Routledge
605 Third Avenue, New York, NY 10158

Routledge is an imprint of the Taylor & Francis Group, an informa business

British Library Cataloguing-in-Publication Data
A catalogue record for this book is available from the British Library

Library of Congress Cataloging-in-Publication Data
Names: Baxter, Dennis, author.
Title: Immersive sound production: a practical guide / authored by Dennis Baxter.
Description: Abingdon, Oxon; New York, NY: Routledge, 2022. |
Includes bibliographical references and index.
Identifiers: LCCN 2021036324 (print) | LCCN 2021036325 (cbook)
Subjects: LCSH: Auditoriums–Electronic sound control. |
Theaters–Electronic sound control. | Sound–Recording and reproducing. |
Entertainment events–Production and direction.
Classification: LCC TK7881.9 .B39 2022 (print) | LCC TK7881.9 (ebook) |
DDC 621.389/7–dc23
LC record available at https://lccn.loc.gov/2021036324
LC ebook record available at https://lccn.loc.gov/2021036325

ISBN: 978-0-367-51220-0 (hbk)
ISBN: 978-0-367-51219-4 (pbk)
ISBN: 978-1-003-05287-6 (ebk)

DOI: 10.4324/9781003052876

Typeset in Bembo
by Newgen Publishing UK

This publication is dedicated to my wife Charlotte Baxter and to those who have fought the battle against cancer and won, and to our grandson Kolebi Dylan Baxter and to the many others that have lost the fight. To my mom who shared music with me at birth and my father who made my first bass guitar and gave me a Chet Atkins record. WOW. Once again I am grateful to my wife Charlotte who has been my lover, editor and soundbabe, and to our son Devin who has been my audio assist and audio manager at many Olympic Games. Deeply, and from the bottom of my imperfect heart, I truly appreciate that my wife and son stuck by me and tried to understand and give me the support and freedom to pursue a dream – several dreams. To Taryn Baxter, who gave my family three beautiful grandchildren – Karli, Kaylei and Kolebi Dylan.

Contents

Figures

Contributors

James M. DeFilippis has been employed in radio and television broadcasting over the past four decades. Currently a consultant to the media and broadcast industry with a focus on UHD TV, wide color systems, file-based production work flows and immersive audio. Some of his accomplishments are: progressive video camera systems, MPEG 2 splicing for network distribution and the first use of an all-digital, disk-based, super slow-motion camera/recording system at the 1996 Atlanta Olympic Games.

Jim received his BSEE and MSEE from the Columbia University School of Engineering. He is a Fellow of the SMPTE and a recipient of the SMPTE David Sarnoff Medal for his work in advancing the art of television engineering. He actively participates in developing standards in the AES, ATSC, SCTE and SMPTE.

Michael "Mike" Kahsnitz was born in 1958 in Germany. After school and studies at SSL, 3M Minicom Division he worked as technical director and sound engineer at Dierks Studios in Germany for more than 20 years. He also founded with partners his own PA company in 1976 and later on the remote recording company Eurosound GmbH and performed hundreds of live recordings with many well-known artists. In 1989 he also became a specialist for audio measurement at Audio Precision with RTW in Cologne. He is still working for RTW as senior director of product management. He worked as a teacher at the school for broadcasting technologies in Nürnberg. He´s a member of the EBU PLoud group and AES and has written various publications and convention speeches.

Preface

Future Entertainment and the Implementation of Next Generation Audio

Immersive and interactive sound is coming to the consumer quicker than expected. 4K televisions are on every electronics retailer's shelf, right along with an easy-to-install immersive sound soundbar speaker system. The industry is ready with standards for immersive audio and high-definition video, there are encoders and processors for distribution, but most importantly, the consumer can experience a better sounding product over soundbars and headphones.

Is it perfect immersive sound? Good question, certainly not. Does it sound better to the consumer? Absolutely. Remember, it will be the consumer who will get the ear of the content producers and drive the adoption of immersive sound.

In many parts of the world sound production for picture and music is being implemented or planned for with immersive sound as the future-proof standard. For the audio practitioner, the question is how to produce sound for the future.

There is little information and minimum practical experience on how to produce immersive sound live and with this book I plan to share my experience along with some proven and tested best practices for producing immersive sound. I can assure you immersive sound is more than a *blackbox* or magic. Sound professionals need to understand the principles and best practices of immersive sound to remain competitive in their field.

This publication is written for the beginner, all the way to the technicians and applications experts. Using case studies, I will examine the technologies that make immersive sound possible for the audio mixer, sound designer and content producer to craft a compelling soundscape. In addition to the hardware and software technology, the immersive sound mixer and practitioner must develop the necessary listening skills.

Immersive sound is the standard for the film industry and is rapidly becoming the audio standard for streaming and broadcast television. Immersive sound is part of the ATSC 3.0 (Advance Television Standards Committee) audio standards for ultra high definition (UHD) broadcasting and was quickly adopted in 2016 by Korea and shortly thereafter by the United States. Additionally, immersive sound is the audio standard for Sky Sports, BT Sports and NHK Japan, who produced and broadcasted the 2020 and 2022 Olympic Games in 8K Super UHD with 22.2 channels of immersive sound.

I hope to clearly present innovative and creative techniques for the capture, creation, compilation and reproduction of immersive sound that will differentiate future content and the next generation of audio professionals from a two-dimensional world.

1 Guide to Future Entertainment Production and Next Generation Audio for Live Sports

This publication compiles my research and search for the perfect enhanced entertainment experience. Future entertainment production harnesses state-of-the-art electronics and emerging technologies with progressive production concepts and practices to prepare for and manage the audio and entertainment landscape into the future. Next Generation Audio was defined and investigated as early as 2012, however the future of engaging entertainment is a work in progress that goes beyond sound and merges the interactive aspects of games, control features of secondary smart devices and enhanced engagement features to lock in an easily distracted consumer.

Sound is all around us and immersive sound can be described as what we actually hear from multiple sources and directions and consisting of a wide range of amplitudes and frequencies. We are bombarded by noise and even controlled sounds, but our sonic perception is further complicated by the collision and reflecting of sounds that create a complex acoustic soup. Fortunately our brain can selectively and subjectively filter sounds to decode our surroundings, but sometimes the brain just wants to rest and be entertained and as sound designers and aural artists our responsibility is to create perfect soundscapes without too much useless information. Electronic media must be designed and produced with the consideration that sound and visual entertainment is consumed under a wide range of attention and engagement.

This book is a functional guide for designers and producers to explore their creativity and ingenuity in the production of next generation sound and advance entertainment features.

Where Did We Come From?

Many advances in sound for film, broadcast and streaming media have centered around transmission and reproduction. The evolution of audio in broadcasting has paralleled the increase in the number of record and transmit channels available in the broadcast ecosystem. Analog satellite technology provided two channels of audio and often one channel was for full program and one channel for program without commentary. Through the early 1980s stereo was often processed from mono and originally surround sound was delivered over a two-channel analog transmission path that was decoded in the home. Analog sound could only expand linearly, but with the digitization of sound advances in technology and application accelerated significantly because manipulating digital sound was easier and cheaper. Digital sound gave us the possibilities of significantly increasing the channel count which resulted in surround, immersive and interactive sound over a very limited bandwidth. Sound and video reproduction is as much about transmitting the signals to a device and then reproducing those images and sounds to an audience as it is about the actual production.

Advances in electronic video have been around bigger and better picture quality with the introduction of high dynamic range, faster frame rates and significantly greater resolution with 4K and 8K video. However, consider that all high-definition pictures are still two-dimensional

DOI: 10.4324/9781003052876-1

Figure 1.1 We are all surrounded by sound. Photo courtesy of Genelec

and depend on the sound to bring expanded dimension and true realism to the experience and picture. Just as surround sound added depth to the horizontal orientation, immersive production adds the vertical orientation and height perspective.

Immersive sound can be defined as the final evolution to a total and complete 360-degree aural experience. So, my question is, will there be another Next Generation Audio definition of immersive sound after this generation of immersive sound has matured and evolved? Most global broadcasters are using 5.1.4, the film industry likes 7.1.4 and the Japanese have developed and support 22.2 channel immersive sound. All formats qualify as immersive sound, so which format is better? Which format reproduces and sounds better over various speaker configurations or over soundbars and headsets?

Future ready entertainment is about production practices and technologies ready for full integration into the next generation of capabilities and potentials and not just about better sound for a bigger picture. Immersive sound is not about what we already have, but about new developments and the implementation of advance audio practices and future resilient technology. Consider the fact that the technologies developed by AURO 3D, Dolby's Atmos and

Fraunhofer's MPEG-H did not invent 3D sound, but these technologies provided the tools for creative sound designers to compose with.

The future of entertainment is supported by skilled sound sculptors and entertainment integrators creating the next experience. Next Generation Audio (NGA) is where we are today, but I don't see things sitting still. Just think about haptic, parallel goggle experiences, hand gestures, interactive storylines and even sound shaping and steering.

There is a growing toolbox of hardware and software along with online support resources and references to guide you to success. Advance audio practices and theory is an umbrella of production technique and technical production practices that provide guidance and a road map for a successful immersive sound production and deployment. This publication is focused on the production and reproduction of future entertainment systems and NGA which is much more than greater channel count for immersive sound and interactive components, but includes intelligent rendering for consumption and distribution as well as a host of menus for additional personalization.

Life was simple when we had analog television in big wooden cabinets with 12-inch speakers and a tweeter, plus high fidelity sound that was delivered over FM radio frequencies. Today's sound reproduction over consumer devices seems like a moving target. Consumer choice includes listening over a wide range of devices and formats including mono, stereo, surround and immersive over speakers, soundbars and headphones.

Some consumers are not interested in the immersive experience but clearly enjoy the interactive aspects of Next Generation Audio. Interactive sound, also known as personalized sound, offers tremendous possibilities for content producers to extend the viewing experience. As broadcasters fight for the attention of a younger audience who expect to be able to control the experience, the user interface is critical to attract and keep new consumers. Imagine the ability to select the radios of your favorite driver and crew while listening to the energy of the sounds trackside. Significantly, content producers can put limits on any interactive parameters such as the range of volume levels or choose alternative content such as different announcers or even develop income opportunities with secondary pay for content that can be controlled.

Interactive options can be executed/implemented in real time and at the fingertips of the audio practitioner and producer in the outside broadcast/remote environment. Personalization will only develop and evolve with the invention and developing of novel technology and unique applications and by monetizing exclusive content.

Future ready sound is producing the sound for whatever the next generation wants. Expanded features and production tools will grow in sophistication as a result of greater demand from consumers for more advanced attributes and benefits, including both entertainment features as well as utilitarian features. With entertainment features, such as choosing to listen to the coach or different commentary, or utilitarian enhancements such as improved speech intelligibility or playback over a wide range of reproduction devices, there is no doubt that user control will be demanded by next-generation consumers.

This book will guide you through the capture, production and reproduction of the dimensional and personalized consumer sound experience and explain how audio practitioners will implement the imminent wave of new productions. Future entertainment systems will continue to evolve because there will be a greater and diverse demand for integrated, immersive, interactive and intelligent entertainment and sound. Perhaps the next phase in creating a truly multi-dimensional experience would include alternative sonic and tactile experiences or further tapping of the brain to convince us to accept different realities.

We are completely surrounded by sound – or the absence of sound. Even though we are endlessly assaulted by sound we live in some level of sonic saturation that is likely some variation sandwiched between. Sound can elicit a wide range of sensations and stimulus that can evoke emotions, provoke involuntary physical reactions or simply be ignored as background noise. Sounds can be scary and rhythmic sounds can be tribal and trance-like accompaniments to dance or religious chants. Sound gives direction and moving sounds can give us warnings. Sound is a powerful weapon and tool where sub-harmonic and ultrasonic sounds are used both in warfare and as tools for diagnosis and cures. Sound is the most basic form of communication and entertainment and has been around us long before the commercialization, computerization and electrification of sound as we know and currently use it.

As audio capture and reproduction has expanded from a single dimension sonic space to a fully immersive, multi-dimensional soundfield, sound designers and audio practitioners face many challenges in practicing and realizing the production potential of multi-dimensional sound. Clearly Next Generation Audio is a paradigm shift with the addition of height sound information completing the illusion of dimensional sonic space.

You could consider that stereo was more dimensional than mono, but surround sound changed the soundscape of electronic media and there is no doubt that immersive sound completes the impression of sonic space. The use of the height speakers is subjective and clearly dependent on the content, but one thing is clear, height speakers give greater resolution and creative possibilities in audio reproduction.

First, consider how the ears hear sound and how the brain interprets the location of sounds. The level of detail and amplitude is at its greatest when the sound is front and center to the listener and decreases in detail and amplitude when a sound is directly behind or above the listener. This creates various levels of emphasis across/around the 360-degree spatial panorama of the listener, but sound localization can be very disconcerting to the listener when sounds do not appear where they are supposed to.

Localization is especially useful to the immersive sound designer and storyteller, but don't wear the listener out because there is clear evidence of sound-induced fatigue from high volume levels, distortion and excessively complex sound content.

Immersive sound for media is a capture and creative process, but ultimately a soundscape is what a sound designer sculpts and wants you to hear. Immersive sound can range from faithful capture and reproduction of the soundscape to cinematic expressionism. Immersive sound is the next level of sonic interpretation and control: aural sculpting for entertainment; creative interpretation; physical and psychological manipulation by sound. Immersive sound is a new tool in storytelling and gives the producer the ability to steer and guide the listener through the soup. Clearly, immersive sound production can be used to more accurately recreate a soundfield or to reconstruct and reinforce the desirable features of a soundfield.

There are varying interpretations of what immersive sound for broadcast and broadband should be because there are at least two distinctly different content genres with specific interpretations and workflows. These are post-produced programming and live entertainment and sports. Each genre benefits differently from immersive sound production. Frankly, we are just beginning to explore these very real possibilities. As with 5.1 surround sound, immersive sound programming will lead the integration and migration of immersive sound into the consumer use-flow/domain.

In post-production there is no doubt that movies and made-for-TV serial and drama programming benefits from immersive sound production. The audio for these media productions is produced in studios and on computer workstations and there are plenty of tools that work within the digital audio workstation (DAW) domain.

Live music and entertainment production will benefit from the enhanced immersive experience and from the creative palette of spatialization tools for innovative imaging. With immersive sound production, the listener enjoys an enhanced spatial and localization experience shaped from spatialization tools and advance audio production techniques which create spatial imaging far beyond any horizontal speaker reproduction.

Certainly lavish soundscapes created in post-production are the clear and obvious bene-factor of the complete immersive vibe, but simple soundfield enhancements create an enter-tainment experience as well. With surround sound, there has always been a question about using the surround channels for specific sound. Now, with immersive sound, the question is what sound do you put above the listener? What do you put in the height channels when there is no apparent height sound source?

Many skeptics look at immersive sound as surround sound with more trouble. There is no doubt that there is going to be a learning curve, but next generation sound offers unparalleled creative possibilities with consistent reproduction benefits to more consumers across a wide range of both fixed and mobile devices.

There are many questions relative to the actual use of immersive sound in broadcast and how immersive sound can best be used to produce an engaging experience and not just fill the sonic space with noise. Immersive sound is defined by a variety of criteria, but clearly immersive sound is the next step in the complete 360-degree sensory experience. Unlike the eye, the ear is capable of detecting sound all the way around us, but what does the brain need and want to hear in this sonically congested world?

Moving from a single point mono reproduction of sound to two channels, six and on to twelve channels, most certainly increases the sense of realism because sound is coming at the listener from 360 degrees. However, stereo and surround reproduce sound on a single horizon with sound in front of, to the sides of and behind the listener; the x and y-axis with what was obviously missing is sound from above the listener – the z-axis. Sounds exist above the listener either directly from sound elements, also known as objects, or indirectly in the form of sound reflections, which makes the real question for sound designers one of how to use the sound above the listener. A holistic 360-degree view is a spatialization approach to sound design while sound localization can be viewed as using a part/section of the soundfield to achieve the effect.

The use of the height speakers is subjective and clearly dependent on the content, but one thing is clear: height speakers give greater resolution and creative possibilities in audio repro-duction. An enhanced sports-style experience does not require precise localization with most height sounds and can be described as an ambiance enhancement. Similarly, using the vertical axis to expand the appearance of space can unmask conflicting sounds.

Future sound production is an evolving bundle of advance audio production techniques as well as technological progress and creativity. Immersive audio requires more channels, but little more transport bandwidth than its 5.1 predecessor. There are high-quality low bitrate encoders and other schemes that can transport Next Generation Audio include everything from 5.1.2 to 22.2, essentially any possible combination of speakers/channels. Advance audio production practices consider discrete height reproduction as well as up-firing soundbars that are capable of creating a height perception. However, advance mixing and DSP processing is no match for true rendering capabilities.

Crafting the Consumer Experience: Immersive Sound Production

This book will examine and present various levels of complexity in the production of immer-sive sound, from using only the natural sound elements that are captured and reproduced to the extremes of a complete manufacture of the soundfield. There are several methods for pro-ducing immersive sound – speaker/channel-based, the use of objects and ambisonics or even a combination of all three methods. Live, real-time production and reproduction tend to be channel-based because there is no rendering or other inductions of latencies.

In addition to monitoring, live sports and entertainment productions demand real-time tools to meter and manage the expanded channel count. The fact is that the tools are different for real-time applications than for monitoring and metering on a DAW in a studio environment. Monitoring and metering immersive sound production is detailed in Chapter 6.

Speaker/channel-based audio production is the way stereo and surround sound is produced and mixed. It is predicated on the assumption that reproduction will be on a similar speaker system and setup; for example, two speakers for stereo and six speakers for surround. This stereo approach was acceptable for a long time because many television sets had enough room for two speakers and people who were interested could hook up their TV and VCR to the stereo system.

Everything was good until surround sound became the new normal. With surround sound, most consumers did not have adequate reproduction systems, or if they did, they were mostly not set up to match the configuration assumed by the audio mixers and artists. Too many wires and too much hassle. Until the arrival of consumer-friendly audio systems and soundbars there was generally no way to hear dimensional broadcast sound. Immersive sound adoption will absolutely depend on the availability of consumer solutions like immersive soundbars and variations on up-firing speakers to accelerate the adoption process.

Soundbars are not intended for critical listening or mixing, but soundbars have resulted in a significant improvement in the sound quality and experience for consumers. You don't have to think too hard about it, but the future of sound listening is probably going to be simple: headphones and soundbars. And this begs the question of which production method – speakers, objects or ambisonics – can deliver a good experience and reproduction in the real world of listening options.

The gaming and film industries adopted object-based audio because it is an efficient and effective method to deliver sound assets with localization and other information. Object-based technology allows for the audio elements to be kept in the bitstream in a session file and positional metadata is then added on top during mixing and then rendered for playback. On the playback side objects are rendered by the DSP independent of the playback system. Future ready sound labors to improve capture methods for live as well as post-production with improved workflow and integration. Solid State Logic (SSL) has a direct interface and support for the Sennheiser Ambio microphone, which streamlines the workflow from capture to advance audio applications in the mixing desk.

Future Entertainment

Future entertainment requires continuous evolution of the tools and resources needed for production. Real-time signal management, metering, monitoring and mixing platforms are critical for live production. The mixing platform not only manages the signal flow and routing but also provides for spatial placement, localization and spatial processing and signal management.

Sound design often dictates microphone selection. For example, VR ties a 360-degree camera with a single focus point 360-degree sound capture. Because most of the sounds needed are immediate, then close ambisonic capture is probably your best choice, but not the only option. Microphone skills will always be necessary because there probably will always be some natural capture, especially in sports. Microphone capture theory is covered well in Chapter 5 and includes practical tips and recommendations for a wide range of applications and situations.

Future sound includes interactive audio that clearly will engage and attract the gaming generation, but the future of sound is also about personalization features way beyond closed captions. Clearly a friendly user interface will be part of the future of the second scene experience, but content, features and ease of use will drive the adoption.

Selectable enhancements, such as individualized audio channels, not only provide new fan experience, but income opportunities. The viewer often sees the coach but rarely hears the coach because of the distance and background noise – plus the colorful language keeps coaches off the air for most real-time over-the-air productions. There are fans that would pay to hear a foul-mouthed coach. Interactive features allow for content selection and control, but content providers still have limits controlled. For example, if advertising is essential then the content providers can restrict the consumer from turning the voice off.

The future of sound is intelligent audio, which is the greatest benefit to consumers since the beginning of sound reproduction. For decades, Hollywood has remixed audio for a wide variety of listening formats because there is no magic process to recreate the proper balance and dynamics of a multichannel immersive soundscape to two television speakers or a soundbar – yet. Part of the downfall of surround sound was backwards compatibility, but future sound technologies can intelligently render audio to the desired format.

Typical sports audio production is surround sound, however general consumption could include anything from built-in television speakers, single unit soundbars or hopefully a more elaborate setup with satellite speakers or up-firing speakers. How do you get a decent sound over so many permutations? There are problems in the downmix with surround and there would inevitably be problems with any immersive sound downmixing scheme. Downmixing began simply with just combining channels. First generation metadata was supposed to improve the situation, but fixed parameters obviously don't work and new methods and technologies were needed.

Rendering audio for specific formats and treatments is a paradigm shift. Consider that there is a repository of sound channels and objects and the user tells the audio engine to render an optimum mix for 5.1.2 or 7.1.4, stereo or any combination of channels from the cache of sounds. Or the user tells the audio engine to render ambisonics with head tracking for binaural consumption. Rendering the sound to the reproduction device decouples the production process from the reproduction. Rendering is capable of reproducing audio to virtually any consumer format including a wide range of speaker formats, soundbars and ear devices.

The sound designer/producer cannot underestimate the impact over and in the ear listening devices have had on consumer and professional applications. Headphones are required for the new realities – virtual reality (VR), augmented reality (AR), mixed reality (MR) and extended reality (ER) – where complete immersion and head tracking are a requirement. Binaural is a superior format for source localization and panning and is found in most DAWs. Binaural audio is an accurate method for creating a dimensional audio environment (3D) by rendering a two-channel audio representation using the familiar head related transfer function (HRTF). This simple setup requires only headphones and considers transaural playback, room acoustics modeling, and enhanced HRTF equalization algorithms.

As sound practitioners, designers and producers our responsibility is to make it happen. I think the Covid-19 crisis of 2020–2021 brought fresh thinking and hopefully a revived vigor for live production. There are many operational and workflow considerations to assess, particularly when some workflows and technologies are still in development and evolving, but accept the fact that new sound realities are here to stay.

Consumers are seeking more advanced entertainment stimulations and sound is a big part of this equation. With previous generations of broadcast sound technology, the improvement in sound from mono to stereo and stereo to surround sound was obvious to consumers. Research now indicates that the addition of height audio provides significant improvement in the listening experience. Immersive is the final ingredient to complete audio captivation and engagement.

First Generation Audio is simply about minimum features and limited bandwidth delivery. Future technologies can be implemented a piece at a time and scaled up or full up. The functionality and the license reside in and with the device. For example, MPEG-H is already entrenched in Samsung devices and authoring software is available online from Fraunhofer. Terrestrial base – Over The Air (OTA) and internet-based Over The Top (OTT) delivery along with a constant wave of new applications are the future. Future-ready entertainment is the complete bridge to the future of electronic media consumption.

2 Enhanced Entertainment Experiences
Sound Design for Sports

Sound design is usually interconnected to the type of media the specific aural landscape model is intended for. Essentially sound design involves the artistic and creative skills of storytelling along with enough technical knowledge and expertise to implement the sound design. After reading this book you should be able to easily sit down and write your sound vision and have the technical savvy to execute your design.

Sound design for moving images intentionally gives more information than is necessary to support the visuals, however I would argue that good sound design will tell the story without pictures. My picture-friends bristle up when I remind them that radio came before television. Think about the brilliant radio dramas of the past, such as "War of the World", where a few simple sound effects, musical notes and the powerful voice of Orson Welles shocked the world into belief that the earth had been invaded by Martians. Undeniably, that's great sound design.

With film, 360 video and the new reality worlds of virtual reality (VR), augmented reality (AR), mixed realities (MR) and extended reality (ER), the majority of sound design is done in post-production, but with live ultra high definition (UHD) broadcasts the sound design is performed in real time where there is a completely different process and workflow. Lavish sound productions with moving sonic objects and hyper-reality sound can easily be accomplished in post-production because the outcomes can be dialed in and auditioned for their effectiveness, but in live sound there is no "do-over" which often results in a sound practitioner being conservative and cautious.

Sonic Intensity vs. Sonic Intimacy

Listeners are often assailed by sounds competing for their attention. Forcing sounds louder just to be heard is not good sound design and this practice ultimately led to different loudness schemes and legislation to get the audio under control. Volume leveling is only one tool, but consider that volume dynamics may be a better way than brick-wall leveling and over compression. Dynamics is the difference between a soft part of a soundscape and the loud parts, which creates variation in the soundfield and interest for the listener. For example, when I mixed motorsports I would momentarily allow the dynamic roar of the motors to swell above the commentator's voice to give a specific sonic interpretation. This resulted in some complaints from the producer about burying the announcer's voice, but I argued that racing is a sport that by its nature is loud and that the audience expects the sound to be dynamic. I truly believed that this sound design enhanced the realism of the event, sport and broadcast production.

Some might say that capturing every detail of an event and venue should be enough to deliver more than the illusion of reality since the sounds are the real deal. Right! Maybe, but sometimes the real sounds are not what the audience is expecting. Blame it on film and games. The best film example is the movie *Dirty Harry*. The gunshots were menacing and the audience knew Harry's gun was lethal because of the bigger-than-life soundtrack, but anyone that

DOI: 10.4324/9781003052876-2

has ever recorded gunshots knows that real gun fire does not really sound like a movie. Plus, it is difficult to capture a sound with such a wide dynamic range and sound pressure level. This is where sound design is at its best. When you add a little thunder, dynamite explosions and Voodoo to the mix, that's the *Dirty Harry* sound, bigger than life.

With film you often create and sculpt the sound in post-production, but what about live?

Live sound can be lackluster and underwhelming, not always because the sounds were not believable, but because some sounds are boring or not available. When I began my initial sound design for Velodrome – track cycling, I sat by the wooden track and listened. Track cycling on a wooden track is very faint and difficult to capture. I decided to go under the track and to my amazement I heard a symphony of sounds that became the heartbeat of my new sound design. I heard a rhythm as the bicycles moved along the track. I heard a steady beat that occurred as the bicycles passed over the seams between the sections of the floor. Clearly this was not the natural sounds of track cycling, but the rhythm was majestic, trance like and interesting. The sounds gave a sense of motion and speed, and became part of my soundtrack of track cycling.

Often there are technical limitations to deliver the sounds that the viewer/listener is expecting. When you have sounds with roughly the same volume level or similar frequency content usually the loudest sound will "mask" the softer sounds. For example, motorsports immeasurably overpowers virtually all the surrounding sounds – particularly the audience. I was frustrated that I could not capture the sound of a very large field of spectators specifically at the finish line. You clearly see the fans' exuberance, but as the pack of cars crossed the finish line the sound intensity of the cars overwhelmed the cheering fans. You know the sound of the crowd was really there, the broadcast audience knows that the sound of the crowd was really there, but you could not capture the intensity of the crowd with typical broadcast microphones.

The solution was obvious: sound supplementation. I began my career in a recording studio and transferred those experiences to the live broadcast productions and began using sound loops and samples from the very beginning. I was told by some of my contemporaries that I was cheating and that using samples was deceiving the audience. I questioned how that was possible when I was using the real sounds and often the real sounds from that venue or sport. I argued that I was satisfying the expectations of my audience and I am not creating a documentary. Who am I cheating? I find it ironic that during the Covid sport season of 2020 my critics were, for the first time, compelled to create sound to cover sound-starved stadiums and satisfy the expectations of fans. Is it live or Memorex?

Sometimes enhancing existing sounds heightens the illusion of reality. When surround sound came along so did the low frequency effect (LFE). The LFE is not the subwoofer or any sort of low frequency management. The "E" means effect – period. I would make a sub-mix of specific microphones, apply a high pass filter, as a result filtering out any frequencies above 100 Hz and send that signal to a sub-harmonic enhancer and then directly to the LFE. I would use the sound sparingly, because the rumble could become obnoxious, but enough would give the viewer the impression of being next to the track. You could feel the sound. See Chapter 9 for more on this.

Off-screen sounds are a staple in film and theatre soundtracks because they provide continuity to the sound mix. For instance, the "birdie microphone" in golf provides the glue to the soundtrack when cutting between different holes and locations on a golf course. You only need a couple of good locations because the non-specific birdie sound should be able to be used at just about any place on a large golf course. The birdie sound is something that must be authentic – just ask a major network that got a call from the Audubon society declaring that the bird samples used at that particular event were not native to the region. Oops – fake birds.

This is an incident where creating new sounds detracted from the story. Remember, your audience will notice if something is missing or wrong.

Film and game sound design often creates sounds that do not exist, where live sound often enhances sounds that really do exist. The sound designer often will create layers and textures of sound to enhance the mood, ambience and tone of a soundtrack. By creating layers, the sound designer can effectively and efficiently balance pieces of sound including commentary, atmosphere, effects and music to find the right blend.

Music is outside the event and often is used to enhance the drama or mood of a feature story, or just signal a commercial break. You might want to consider that sounds and tones could be used as the heartbeat of your sound design and the rhythm of sounds can drive a soundtrack. Simple synthesizer enhancements can be very effective and not distracting. Future ready sound design will embrace the entertainment aspects of an event and I can absolutely declare that spectacle will rival the sport in the future productions and entertainment designs.

There is no doubt that sound designers will inevitably collaborate with the picture producer, but creative designers may have a sonic vision in their head before they start. It is a good idea to feed the producer ideas since their experience and vocabulary may be inadequate to express their true ambitions.

Recreating the height aspect in a 3D mix is challenging when you have different practices and technologies reproducing height to the consumer with different results. Undeniably sound design is part art, science and creative magic, but should begin with a vision and a set of achievable goals. Production sound design is the creative intent, vision and roadmap for the sound producer. Production sound design is the personal influence a sound artist leaves on a soundscape who ultimately reproduces the experience, enhances the experience, or possibly even alters the experience of the listener. From pure reproduction to pure fantasy, the sound designer may want to create the illusion of the viewer being there or perhaps enhance the audio experience of the viewer/listener by giving a hyper-realistic aural interpretation.

The quest for dimensional sound for radio and television began with the journey from mono to stereo and as immersive sound ensconces itself with consumers there are even advocates for channels formats all the way up to 22.2. The ongoing advances in audio for electronic picture have been significant benchmarks in the enduring pursuit for sound that enhances the believability and reality of a two-dimensional picture.

The future of sound production has undoubtedly change because current re-production practices for stereo and surround is flawed. Dimensional sound's nemesis has been the expectation that the reproduction setup will have the same number of speakers and prescribed location as the production. Channel/speaker-based production has been the weak link since the introduction of stereo sound. Particularly now with soundbars, you have to wonder if your mix is sounding like what you are crafting or a much less optimum sound experience.

Even before sound became part of the movies, a live musician could resonate an emotional chord and suspend any skepticism that the audience may have looking at the bigger-than-life screen. Sound is essential to the believability and the sensation of reality. You know a good sound design when you can close your eyes and mentally touch and become a part of the picture.

For a creative sound designer, immersive audio will draw you in because of the cool possibilities to manipulate more sonic space. But after a while, you begin to understand the utilitarian benefits of this new sound paradigm. Each genre of media production benefits differently from immersive sound production, obviously resulting in a variety of sound designs and audio objectives. Now the challenge for the professional audio community is to implement effective and practical production and reproduction practices for a scalable sound design that supports the increased channel count, interactive and intelligent features.

One obvious and significant addition to Next Generation Audio is the aspect of height information. The obvious use of height would be to deliver what natural sounds are really above the listener. Maybe a not so obvious use of the height aspect would be to create sonic landscapes, such as full immersion underwater or the sonic vacuum in space. Immersive sound has been researched and documented to be a key component to the listening experience, but immersive sound may not be the end result of a particular sound design.

A sound design begins with a sonic vision followed by a solutions analysis. By this I mean you can dream a wonderful sonic landscape that is impossible or impractical to deliver. Next Generation Sound has an abundance of possibilities and solutions for personalization and interactive entertainment experiences and sound designs and not as many pitfalls as previous sound advances and solutions.

Personalization is a creative as well as utilitarian tool. For example, a user could pick their announcers and language while people with hearing difficulties or difficult listening environments would benefit from the utility aspect of dialog selection and level controls. Personalization doesn't mean the same thing to different people, particularly when it comes to entertainment. For example, increased channel count can be used for creating immersive soundfields, but some consumers are not interested in dimensional sound. Creative personalization is content driven and user centric. For example, a program that has multiple dominant sound sources and speech intelligibility is critical such as with a split screen of a shared house. Think *Big Brother*: this programing would benefit from sound selection, localization and isolation in the boundaries of the space and an observer could dial into multiple conversations with clarity and without confusion.

Immersive and interactive sound design can range from the basic to much more advanced sound enhancement, but I assure you advance channel count and interactive is a paradigm shift. The forward-thinking sound producer should consider future ready sound as complementary production differentiators in advance sound design and that advance sound design is a competitive advantage to set your productions apart.

What Is Immersive Sound for Live Sports and Media Production?

What is the sports fan expecting from immersive sound or other interactive features from an event, sports or venue experience? What sounds should be in front of and above the listener? If you are listening from a position above the field, what is the sound that you hear and is this sound beneficial to your sound mix? Now consider using the height channels to define and enhance the perspective. Do you add more crowd to your soundfield? What do you let the fan control – balance, alternative channels? What?

Sound Expectations – Replicate or Recreate

Sports sound requires the mixer to reproduce aspects of the experience because there is a certain expectation that is ingrained into the viewer. Our brains have a very long memory for sound and what something should sound like. The sound of the corner kick in football, the swish of the basketball net or the crack of the bat are the unmistakable and expected sounds of sports. These examples are natural and organic, but what about sound enhancements?

For instance, darts is a popular sport in the UK with a cozy but interesting soundfield. There is a distinct sound when the dart impacts the corkboard, but there have been rumors circulating for some time about sound practitioners using the microphone on the backboard to trigger additional samples to enhance the impact. Subtle enhancements from explosions or thunder would certainly heighten the experience but a few questions arise. Is it necessary? Are there any ethics issues from a deceptive practice?[1]

The concept of immersive sound attempts to put the listener in a complete soundspace with sound above and around the listener as is heard in the real world. The principles of immersive sound production provide methods for the sound designer to produce a dimensional soundfield for the listener to experience, but requires new ideas and advanced tools.

The main differences between immersive sound in live broadcasting and immersive sound in post-production, is the difference between "channel-based immersive" and "object-based immersive". Channel-based immersive means that most of the immersive work is done live in the mixing desk, whereas object-based immersive uses the full immersive production toolkit and generates metadata for individual audio objects that are included and rendered in the final product. Object-based technology allows for the audio elements to stay in the bitstream along with relevant metadata, although the positional data can be modified and added on top of the audio before encoding for rendering. Live broadcast mainly uses channel-based immersive because it is real time and there is no rendering. There is no time for "do-overs" in live TV.

Where Do You Start?

As I previously said, a sound design begins with a sonic vision followed by a solutions analysis. A sound vision should be appropriate for the media but, remember with the exception of radio, the sound is there to support and, in some cases, drive the picture.

What are the sound designer's goals? These can be stated in broad terms, such as to create a dimensional experience, or more specific, such as to create a wider sweet spot or improve intelligibility. A simple sports sound design might include a dimensional aural landscape with a man-in-the-stands perspective which is often a simple and static view from the side lines. Basic aural enhancements such as widely spaced microphones for the height channels may be sufficient for an improved immersive enhancement.

A design goal to widen the sweet spot in a prime listening space would be desirable for several viewers/listeners to enjoy a program together. Content producers forget the fact that television can be a social activity and unlike a film which often requires full attention, television sports require much less attention.

Beyond the creative possibilities of height production, immersive sound offers some interesting possibilities for improving speech intelligibility. The concept of frontal soundfield reinforcement uses the front height sound projection from the left height channel and the right height channel in conjunction with the center channel form a triad of cohesive sound focusing the sound wave, thus improving the clarity of the speech relative to other soundfield elements. See 'frontal soundfield reinforcement', discussed later in this chapter.

Speech intelligibility is a difficult problem for sound designers because of listener variables including poor reproduction devices and inconsistency in the hearing ability of many listeners. However, user controls are already available and content producers can unlock features where speech channels and other audio dynamics and options of the soundtrack can be adjusted by the listener. I will never forget a listener telling me that they disliked the commentator, but disliked my mix more because they could not hear the commentator clearly. Oops. Speech intelligibility will certainly benefit from consumer personalization and interactivity.

A sound design can include the goals and ideas of more than one person involved in the production. The executive producer's goal of interactivity may be different from an engineering goal of flexible and scalable emission while the audio producer's goal may be to produce an integrated audio experience using multiple reproduction formats such as UHD and VR.

By definition, immersive sound replicates sound above the listener and if the sound designer is crafting an immersive mix with height elements the basics question is what are you using the height channels for and how does immersive contribute to the soundfield? Film and drama have an obvious use for sound above the viewer because the height channels contribute to the authenticity of a production. Clearly aircraft, weather or even birds are really above the ground and contribute to the believability and continuity of the soundscape even though you may not see what is above.

However, with sports it is questionable what useable sounds exist above the spectators and how they contribute to the sound design and soundfield. I have heard 3D recordings of sports stadiums and understand the concept of being at the event, but most microphone methods do not adequately capture and present an immersive sound experience from some mystical best seat in the house.

After the question of height usage is defined comes the question about how the sound is being replicated. A single unit up-firing or side-firing soundbar is not going to be able to replicate a complex aerial scene, but the same soundbar would sound great for a concert with predominantly a front perspective and a massive concert hall ambient soundfield.

The level of effectiveness and aural perception of a sound production should guide the sound designer in the effectual utilization of an immersive sound design. Aural perception is the ability to hear, interpret and organize sounds often based on the attention, inattention and the experiences of the listener. But remember, this can be complicated by technology and for physiological reasons.

The question of effective speaker/channel application applies to all replication sources. For example, surround sound began with simplistic productions just as immersive sound production begins with simple effective production principles. Simple sound design can be as basic as spatial separation of sounds because this effective production method can improve the clarity of the individual sound elements by placing sounds left and right of center where dialog is generally located. Spatial panning is not only effective but, often common in sound design, even using the surround channels for specific sounds can successfully enhance a sound design as long as their use is not distracting.

Irrelevant and distracting sounds are not typically elements of an effective sound design, however audio adventurism can find interesting sound enhancements. For example, the Olympic sound coverage of boxing placed 50 percent of the glove impact sounds in the surround speakers in an attempt to place the viewer in the boxing ring. The Olympics has quality control listeners who are usually well-qualified audio practitioners; during London 2012 one of the listeners called me to tell me there were glove sounds in the rear channels. I told him that this was intentional and a part of the sound design and he quickly quipped that that was not proper surround sound design and that there should not be any sports-specific sound in the surround channels. This is a subjective opinion, however as the sound designer I felt the additional glove sounds were additive in the replication and were not distracting. I also placed the coaches' sound in the rear channels to give the impression of the athlete being surrounded.

The Big Picture

Stereo and surround sound reproduce audio in the horizontal plane – in front, to the sides and behind the listener, while immersive sound finally gives the third dimension – above the listener. The height channels clearly contribute to the desired effect of total immersion by surrounding the listener. Essentially, spatial imaging and placement of sound elements is the foundation to any dimensional sound production and localization in the horizontal and vertical soundspace is essential to realistically creating an accurate immersive soundfield.

Immersive sound can be produced by constructing an ear-level lower surround soundfield and another separate upper soundfield and conjoining the layers electronically or psycho-acoustically to create a holistic soundfield. As with all immersive sound production the challenge is to achieve a sense of continuous dimensional space – sound that is fully contiguous in the complete sound hemisphere.

Spatializing Sound Elements/Objects

As with all sound designs and sound designers, there is a subjective interpretation of the soundscape/images which results in some liberty with the creativity of the soundscape. However, it cannot be overstated that just producing cool and interesting sounds and missing the fundamentals like the viewer not understanding what people on screen are saying is not going to win any awards. General immersive sound design embraces the principles of hearing and sound propagation. For example, our perception of the level of detail of sounds increases from a minimum of detail when a sound is far behind the listener to a maximum of detail when the sound is front and center to the listener. This creates various levels of emphasis across 220 degrees, from left-mid to right-mid in front of the listener, and is a useful principle for object placement in sound design.

Just as the details of the sound increase as you focus forward, the confusion of sounds rises as the sound image increases in elevation above the listener. There is a sound replication principle used in the immersive format AURO-3D called the Voice-of-God which physically replicates sound directly above the listener. Immersive sound design with virtual replication and spatial imaging does not achieve the same effect as real speaker propagation. Sounds above the listener are often confusing but have interesting production possibilities.

Film effectively uses sounds above the listener by building complex sound fields through orientating and spatializing a group of objects in a soundspace above a listener and processing them together. Live production benefits from enhanced spatial production practices that provide effective sonic enrichments over a variety of consumer devices.

Believability: Authenticity vs. Entertainment

Sound designers excel when they appropriately satisfy the expectations of the listener and then take the experience to another level. Immersive sound is a powerful new creative tool that has come along at the right time. The enhancements from the ability to reproduce sound above the listener are significant and clearly provide entertainment as well as functional possibilities and opportunities. For example, you could separate the field-of-play sounds from the atmospheric sound using dimensional sound production and reproduction techniques.

Often mono and stereo sound components and groups are competing for the same soundspace because each speaker/channel contains sound elements from the specific sound group as well as sounds from the ambient sound group. With surround sound it is common to reduce the atmospheric/ambient sounds in the front channels where the specific sounds are usually found. Certainly, atmospheric boost and balance can come independently from the rear surround channels. Immersive sound not only offers soundspace behind the listener but also soundspace above the listener to improve the sonic aspects of the soundfield.

Point-of-View (POV)

One of the many facets a sound designer can control is localization and perspective. Perspective/point-of-view (POV) is the foundation for literary scripting as well as electronic media production. For example, sound design in most visual media environment is presented from the perspective of the spectator, the participant or both. Establishing a point of view is recognized from the very beginning of a sound presentation, however changing the point of view is possible and defendable.

I have heard differing views on changing the perspective of the sound when the perspective of the picture changes. A sports production is a combination of wide and close visuals and the sound perspective can change with the size and content of the picture. For example, Formula One motorsports completely changed the soundscape and perspective when the camera view is from inside the car as opposed to a grandstand shot. Additionally, there is a significant sound shift from a wide camera shot to a camera next to the track. This sound design results in three distinctly different soundscapes: wide car sound; close car sound; and in-car sound, which is consistently changing with the camera cut rotation.

Presenting the sound from the point of view of the spectator or from the point of view of the participant is a significant consideration and is the starting point for any two-dimensional or three-dimensional sound design. Significant visual perspective changes offer the opportunity for more aggressive use of the height channel, such as full immersion when the camera is inside the car. Contrary to the radical changes in picture like with the production of motorsports, field sports such as Premier League Football or professional tennis tend to remain consistent with the soundfield orientation. The effects sounds highlight the action while the atmosphere swells with the crowd reactions, but neither the effects nor atmosphere are enough for a complete perspective change.

The challenge is how to use the additional dimension channels to contribute to the sound design and enhance the perspective of the production. Generally, the surround channels do not change unless there is a specific audio effect the sound designer is trying to achieve – like being inside a racecar as opposed to being in the stands. These types of close-up shots are a significant change in POV and perspective which warrant such a change.

Sport is often presented from the POV of the spectator in the stands with the sound design generally presented with everything in front of the viewer/listener, no matter how big the soundfield. The natural sound of the audiences is generally around you with diffused ambiance and perhaps the PA above the listener. Realistically, the sound from above would be diffused crowd and PA noise and any field or venue-specific sound would be masked.

Sports audio production can be categorized as a generally static perspective or a variable perspective. For example, the perspective of a football match does not change even when the camera perspective changes from a wide overview to close handheld field camera – a static perspective. Only the close-up sound from that specific camera is added to the mix, but the perspective stays the same.

Field sports such as football, baseball and court sports like tennis, badminton and table tennis are presented from the perspective of the person in the stands viewing the competition in front of them. This sideline POV was an easy adaptation for 5.1 Surround Sound because the field of play sports sounds or event-specific sound were in front, next to the screen, keeping the viewers' attention forward. Generally, the surround channels were used only for ambiance and crowd.

With a variable perspective, when there is a significant shift in the size of the scene there often is a change in the scene sound usually with a shift in magnitude and spatial characteristics. This is evident when the sound makes a dramatic change when cutting in and out of the driver's POV from inside the racecar to outside the car. Perspective and point of view are the most natural and powerful tools for the sound designer.

The POV of the athlete/participant is an attempt to place the viewer closer to the competition. For sports, the sound of the participant is up-close audio of the athlete, apparatus and coach. This close perspective and use of the surround channels to include specific sound enhancements brings the viewer closer to the POV of the action. For example, the height sound can be used to contribute to the detail of a soundfield of the coaches' encouragement or sitting in the driver's seat of the car and not just contributing ambiance and atmosphere.

Placing sounds even closer to the viewer gives the impression of the viewer being the participant.

Perception is basic to any sound design and certainly can be enhanced by the magnitude of the sound design and its reproduction. Advanced sound design methods use the surround channels and the height channels to reinforce the competition sound by moving some competition sounds to the sides, middle and above the listener, therefore giving space, separation and detail to the soundfield. Spatialization techniques that use ambisonics can offer algorithms to expand the size of a sound object, thus the perception magnitude of that sound.

Sound Design: Localization

Localizing the sports effects, music performance and voices in front of the listener is a more traditional approach to sound design that establishes the viewer's POV from the venue seating while watching the event in front of them. This puts the sports sound effects in front of the listener and the ambiance sounds around and perhaps above the listener. There could be a sound element or objects that move through the sound field above the listener, however these sounds from above tend to be diffused with an occasional bit of definition.

Not only can the sound designer localize the sound anywhere in the soundfield, the sound designer can control the perception of size of the sound element. For example, sports effects tend to stay relatively stationary in the sound mix and are often accentuated according to the sport and competition. The whack of the bat, splash of the oars or conversation of the Curlers often tell the story and these sounds tend to focus the viewer's attention forward with additional detail.

Another approach to sound design is Camera Cut, a POV where the perception of size of the sound element corresponds to the size of the picture. Typically, a wide shot would have less detail in the soundfield than a close-up shot, but often sports audio production is consistent with a high amount of field-of-play (FOP) sound through most of the camera cuts and visual changes. Sport tends to give a constant spatial audio perspective with medium and wide shots, but can change dramatically with a close-up shot such as a handheld camera. For example, when you show the bobsleigh course from inside the bobsleigh, the intense sound from above, beside, and fully around the listener completes the illusion of being inside the sleigh.

Each perspective appreciably contributes to the believability of the visuals and enhancement of the experience and each perspective benefits differently with the use of height channels. Sound design is an evolving art and understanding how the height channel contributes to perspective is an ongoing process with discussion, mixing and listening tests.

Capture or Create? Audio Production for Live Sports and Entertainment

Live sound production is not always pure capture and live immersive audio production is a multi-step process of capture, create and construct. Immersive sound capture requires consideration of how to acquire a realistic representation of the acoustic space and how the object resonates in that sonic space. Microphone selection and placement contribute to the localization of the direct captured sound as well as the spatialization of that original captured sound. Spatialization occurs in the horizontal as well as the vertical dimension and with microphones positioned along the axis the sounds moving between microphones contribute to a sense of speed and motion.

Each audio element – atmosphere, effects and music – benefits from the additional height dimension which clearly improves the authenticity of the production by expanding the appearance of dimension. There are elements in the soundfield that will need to be created and finally constructed with spatialization into an appropriate mix and format. Sound elements that are stored and electronically reproduced, like sound effects and music, definitely benefit from spatial enhancements and need to be artificially enriched through processing.

In addition to good microphone capture, spatializer tools give the sound designer/mixer the means with which to accurately position the apparent location of a sound anywhere in the soundsphere. Three-dimensional panner/positioner tools can be applied to audio elements or groups of sounds to position any 3D coordinate/position in the acoustic soundscape – around, above and below the listener.

Additional spatial enhancement for ambiance and atmosphere may not be as desirable as soundfield reinforcement. I suggest a positive soundfield reproduction approach to immersive sound design which uses the height channels/speakers to reinforce the soundfield for clarity, detail and sonic quality. Each audio element or group of elements contributes to spatial imaging when captured or created.

Production Philosophy: Atmospheric Enhancement

Immersive sound and the use of the height channels is a creative and utilitarian tool that can enhance the listening experience and dramatically alter and affect the soundfield. There is clear evidence that the addition of sound above the listener improves all aspects of the sound experience including clarity and spaciousness.

Ambiance enhanced sonic experiences are intended to replicate the "being there" experience that is sometimes difficult to convey with the small-screen broadcast/broadband production. Ambiance and atmosphere are the sonic foundation for all formats and must simulate a wide range of visual representation: wide shots, medium shots or close-up shots with soundscapes in stereo, surround and immersive sound. Immersive sound production is achieved by adding sonic information above (and below) the listener. The question is, what level of height reproduction makes a suitable immersive sound mix? What is the content of that height information? With sports and entertainment much of any height information is non-specific and atmospheric.

If you take a literal approach to sound design, you might consider that what is above the spectators/viewers is mostly reverberant atmosphere and PA, but there may be a natural audio discrimination between background and foreground sounds to lessen the audio fatigue. Then you may wonder about its need for immersive sound. Clearly the sound envelope is significantly increased and the listener experience is enhanced, but is this a significant enough enrichment to influence positive consumer behavior for immersive sound?

Fundamental to the auditory experience is judging how the soundspace injects the listener into the visual space. A wide shot of a stadium arguably calls for an expansive, wide sound of the location, but most outdoor open spaces have non-specific diffused sound envelopes that do not contribute any positive acoustic attributes. With immersive sound atmosphere, there is a natural perception that the sound appears more diffused as you move away from the sound source. Most atmosphere-based sound design adds little to the soundfield with spatial clues and often dilutes the detail, masks certain frequencies and often adds a room tone drone quality to the base sound layer. Alternatively, the practice of capturing stratums of sounds that are close, moderately diffused and widely diffused from a soundfield and layering those different stratums of sounds from low to high in the immersive sound mix allows for the mixer to find the right balance between present and diffused sound.

Replicating the event and venue by putting you in the location is a legitimate sound goal. Now the question is what sounds that includes. And not just the obvious. Baseball has a unique venue sound that is accentuated by the organ accompaniments, plus there is a clearly identifiable Hammond organ sound. The organ leads the fans in song and cheer and there is a definite balance between direct and diffused sound. The slap of the ball and glove, the crack of the bat, the umpire calls and the sound of the organ are mandatory in the sound of baseball. College sports is a unique presentation because there is a large band and student body presence. A simple quad microphone array would fill the four height channels with accurate, interesting and entertaining immersive sound. The venue PA is necessary and often too loud and an irritation, but the PA also gives a sound of authenticity and is often necessary for the pomp and circumstance aspects of a media production.

Sport is a televised activity and I question the necessity of extensive atmosphere enhancements for immersive sound production. Consider this: I have heard a couple of sports demos where it sounds like you are in the stadium, but I do not know if more ambiance and atmosphere is compelling enough to move people to immersive sound. Additionally, more ambiance and atmosphere reproduced by an immersive soundbar may dilute any detail in the sound production. Additional spatial enhancement for ambiance and atmosphere may not be as desirable as soundfield reinforcement where I suggest a positive soundfield reproduction approach to immersive sound design which uses the height channels/speakers to reinforce the soundfield for clarity, detail and sonic quality. This may be appropriate for the Hammond organ and the college band immersive sound production.

The height channels contribute to the creation of acoustic space. They reproduce the ambiance above the listener and are a factor in injecting a more spacious impression of the soundfield. For example, the apparent size of an acoustic space (such as an indoor swimming pool) is more realistically conveyed with immersive sound because the acoustics and reflections of the space accentuate what the eyes see and the brain remembers.

Production Note

Using the height channel to create and contribute to the positive aspects of the soundfield is a more progressive sound design intended to improve the sonic quality of a soundfield. This concept uses all channels, including the height channels, to positively reinforce speech, music and effects improving clarity, detail and even speech intelligibility.

For indoor venues and partially roofed outdoor venues, the question exists about how useable the sounds above the listener really are. I have experienced situations where the sound above the spectators differs from what sounds you would expect or hope for. Mechanical noise and sound buildup are a consequence of poor acoustics and will impact a sound mixer's effort in finding, capturing and creating the ambient sound of an event.

Effects Production

The event-specific sounds tend to be in the front of the listener, with surround and immersive sounds supporting the venue experience. To improve detail in the capture of a soundfield, spot microphones are placed in the proximity of the desired sound and the sound mixer includes these close sounds in the mix at the appropriate time, location and volume during the live production. Much of the detail of an audio production is dynamically captured and blended by the audio mixer to enhance the pictures. With the event-specific microphones constantly changing in a mix, the importance of a high-fidelity sonic foundation providing the continuity and glue to construct the complex soundfield on top of, is critical.

Spatializer Tools

Dimensional sound reproduction of spaces is a powerful tool for spatialization and spatial enhancements. Spatial placement is how sounds fit into a dimensional soundfield. Creating sonic spaces can be accomplished with a minimum number of sound sources. For example, processing a sound source into ambisonics not only expands the sound image of the dominant sound source but also dimensionalizes the accompanying soundfield. Spatialization tools can accurately localize sound in a dimensional space because of the high resolution of the phantom images between channels and around the listener.

Creating an immersive experience using sports-specific sounds can be achieved with elevation that is applied to static sound elements or dynamically moving sound elements. Taking an effects mix and elevating and separating the channels has the effect of widening the sonic space and creating a cinematic effect with more sound all around the listener.

Spatialization software can usually be inserted and applied at the channel, bus and group level with insert points, either physical or virtual as in the case of plug-ins. In addition to panning and placement, spatialization tools are capable of an increasing number of tools. THX/Qualcomm has two interesting channel features: distance and size functions.

Distance

Distance is interesting because you are not just changing the volume when you move a sound element closer or farther away, but as in the real world the change in distance can change the tone of a sound as well. Size allows the sound designer/mixer to completely expand or contract the sound element/object like blowing up a balloon – full 360-degree expansion. Expanding the size of an audio element expands the appearance of magnitude of the audio element/audio object by expanding the sound of that audio element into the adjacent soundspace.

Music

Music playback is common in sports production and an element of the production that benefits from immersive treatment. Music can be easily up-processed into the height channels and used as an effective audio element that ties the lower and upper layers together. Spatialization of music uses elevation and panorama to enhance the width of the listener's soundfield and give some separation to the sounds.

L–C–R: Left–Center–Right

There seems to be some usefulness in expanding the center channel from a 5.1 production into the adjacent left–right channels. It seems this divergence of the center channel has become more popular with advertisers and among some types of programming and even with some global broadcasters. There is no doubt that speech in the L-C-R channels improves intelligibility and even using the height channels can also improve speech intelligibility. Note: one advantage of MPEG-H advance audio transmission is that the voices can be treated as audio objects/elements and injected into the personalized mix of that listener and not baked into the mix which affects all listeners.

Sweet Spot

One of the problems with surround sound is that there is normally a confined space that would accurately reproduce all channels of a surround sound mix – commonly known as the sweet spot. Another potential usefulness of immersive sound seems to be the possibility of expanding the size of the reproduction zone – sweet spot. Sound producers are just starting to understand how soundbar reproduction impacts the listener zone. Audio engineers are trying different mixing and production methods and discovering what contributes most to the sound design objectives as well as to the final sound production. Manufacturers are preparing a variety of advance reproduction technologies using speakers, soundbars and up-firing speaker configurations. New transmission technologies have the channel count to deliver the immersive experience. But consumers are on the fence. It seems that while the audio world may be ready for immersive audio, the consumer may not be.

Dynamic sound elements/objects are audio sources that can be panned in the XYZ axis in real time by the audio mixer. This is a very powerful tool that realistically reproduces the sensation of motion. See Chapter 10 for more on this.

Advance Sound Design Practices using Triggered Sound Effect and Samplers

As concepts and practices for immersive sound design and mixing progress, audio practitioners should consider alternative sound designs. Essentially most sound designs are inherently trying to create a sense of authenticity, dimensionality and perspective to bring life to the picture. The 2020 pandemic brought interesting times for live sports broadcast production and perhaps changed the future of live sports presentations. Sensationalist journalism terms, such as "fake", create a very negative description; using the term "fake sound" has nefarious overtones. I believe sound manipulation for malicious interference has ethical implications. But sound supplementation for the sake of enhancing the entertainment value has always been fair game to me. Sound design should boldly embrace sound supplementation as the art of live Foley, which can include crowd-, sport- and venue-specific sound.

Real-time sound supplementation was controversial long before I began using a crowd loop in the early 1980s at NASCAR races. There were always rumors of "birdie" loops at golf events but there was not much of an urgency to "sweeten" golf sound because major networks had the budget to adequately capture the event live. I spent a decade as a freelancer covering Indy car racing and was unhappy that the network crew had microphones mounted on the pit wall at every pit zone while I depended on the sound from two handheld cameras to cover up to 30 pit areas. Why? It costs extra money for extra sound.

Motorsports present unique sound challenges because of the excessive noise and compact design of the track and grandstands. For me, sound enhancement began at the Bristol Motor Speedway where the sound was so intense it made your head hurt. Frustrated with the inability to pull much crowd reaction sound, I built several crowd "loops" to enhance the sound of the start, restart and finish of the race. I knew the real-time sounds were there, but the laws of physics prevailed. It is called *masking*.

Tape loops of sound effects were the shortcomings of sound supplementation because of the inevitable repetition of sounds, but when the music industry introduced electronic samplers that could store real sound and could be played back with a keyboard, everything changed. With instantaneous triggering and access to multiple samples at the same time, the possibility of realistically playing back a complete pit stop from any pit without additional microphones became a reality. What was really cool was adding sounds that were really there but that you never heard before because of masking – like the air jacks lifting the car and two gear shifts out of the pit. I am not trying to persuade anyone of anything, I am only trying to enhance the entertainment value of the content. The sound of air wrenches and engines revving is really happening. I even had different samples of motor sounds for different engine manufacturers.

Let's examine some of the arguments. Question: Are we documenting an event versus presenting a game for the sake of entertainment? Answer: If the intent is to show the negative impact of the Covid-19 virus on sporting events, then quiet empty venues may be appropriate. But if the goal is entertainment for the listening/viewing audience, then sound supplementation is an applicable solution.

The next question is: Does sound supplementation enhance the entertainment experience or is the sound a distraction? Answer: To my ears, poor sound supplementation is a distraction. I spent from July to October 2020 listening to Premier League Football, Major League Baseball, NBA, Ice Hockey, NASCAR and Women's Basketball produced by ESPN. Some sound production is a distraction. For example, my opinion of the sound enhancements for Premier League Football was a half-hearted attempt at sound enhancement with virtually no dynamic articulation. At the opposite extreme was baseball, where they worked with Sony Entertainment in San Diego to deliver a not only very believable, but also entertaining, soundscape for the game.

I talked with Kurt Kellenberger, head sound designer and supervisor at Sony in San Diego. Kurt is a meticulous and innovative sound designer and when he undertook the design for a baseball game for Sony, he studied the broadcasters and picked the best aspects of the game as the baseline for his game sound design. A decade later he provided the sound effects components for the primetime presentation for the Covid comeback of Major League Baseball from San Diego. Sony provided approximately 70 different sound samples and used Abelton Live software to construct a 4.0 sports sound bed for the game. What I heard was not only convincing, but if you closed your eyes, the soundtrack was close to a perfect presentation of the sound – not too much, but interestingly complete.

Faking the sound as opposed to enhancing the sound for entertainment purposes are completely different sound design concepts. There are technical and financial reasons that make it impossible or impractical to capture the desired sounds in a venue or at an event. Capturing the pit sound was a financial decision and only the largest Indy Car event of the year warranted the additional cost. Even though the budget cuts were made clear to the director and producer, they obviously remembered the sound of the Indianapolis 500 and the experience of hearing the sound from every single pit.

To amplify, enhance and even recreate the sound that already exists is obviously different than creating a soundtrack from scratch, which is what a sound designer does with a film or game soundtrack. For example, when the roof camera shoots a pit stop there is a great distance between the camera's microphone and the sound. Even though the sound of the pit stop still exists, the microphones on the roof camera cannot capture the sound. This scenario results in no pit-specific sound even though the listener knows and expects to hear the pit sounds from a distant camera shot.

Baseball deals with more subtle sound plus the concept of home and away teams. Not every team or player is popular and not all the crowd reaction is positive, but what if artificial sound is used for reprehensible reasons such as real-time mood manipulation. As you can imagine, a relentless crowd booing or heckling could result in increasing tension on the field. Dr. Jean Baudrillard wrote about how media affects our perception of reality and how he believed that people often live in the realm of hyper reality as they are connected deeper and deeper to their television, movies, games and virtual reality. Personally, I suspect that at some point numbness sets in and listeners tune out.[2]

Next Generation Sound Design perhaps will be different. Ed Stoltis (A1 CBS Golf) in 2020 commented that listeners will adapt to a new norm, particularly when it is a pleasant and appropriate soundtrack. I think the sound of CBS golf is wonderful with little crowds and no ice machines. ESPN's production of Women's Basketball (WNBA) was appropriate for the

Figure 2.1 Microphone capture from across the pits

picture. The space was compressed with no room for spectators and the soundtrack was the coach, players and commentators. The production was engaging and entertaining. I really like the sound of a quarter-full college stadium and always thought too many drunk fans ruined the TV sound of college football.

ESPN was able to come up with a clever presentation, but large venues used by sports such as baseball, football and professional basketball have noticeably vacant space and seating for spectators who are obviously not there. Directors will adjust with tighter camera perspectives – but the venues are obviously empty. Who is faking whom?

Sound supplementation is nothing new – particularly in certain sports. For example, the Swiss and Finnish sound teams have used samplers for years to cover ski sounds for downhill and cross-country events. Samplers not only fill in the gaps when there are very long camera lenses and few microphones, but when well-executed the additional sound brings the viewing listener closer into the sport and event.

I narrated a BBC radio documentary produced by Peregrine Andrews, titled *The Sound of Sports*. One comment I made that became sensationalized was about using samplers to augment the sound at the Olympics. Somehow that comment got twisted to imply that I was using samplers to fake the sound of most sports at the Olympics. This interpretation of sound supplementation escalated in 2012 when I was accused of "faking" the sound of the London Olympics and was even spoofed by late-night comedian, Stephen Colbert, about the possibility of extremes. You can find it on the web.[3]

Entertainment vs. Documentation

The Covid years were an interesting time for live sound designers, mixers and producers. Spectator-less stadiums are a first and there have been novel ways to compensate for the visuals with goofy-looking cutouts. But usually in sports the sound is supposed to support the picture and what if you are documenting an empty venue? What should the sound be? Clearly, spectator-less venues is a different paradigm that broadcasters and listeners were not expecting.

But I have to ask this question: Are we faking the sound or shaping the sound? The sound samples used in baseball are pristine, probably better than real capture. Is this fake sound? I think some journalists and wordmongers who write about audio know little about sound and use these types of words to stoke emotions on one side or the other on the use of extra sound. I was accused once of cheating, but argued that if I did not deliver a high level of entertainment to my listeners, the only person I was cheating was the listener.

Fan appreciation and participation is often encouraged. Thunder sticks are common at Japanese baseball games, Brazils fans interact with drum corps and at football matches one could never forget the 2010 World Cup in South Africa and the menacing vuvuzelas.

The vuvuzelas dominated the soundscape and aggravated television audiences around the world to the point that within days of the start of matches there were vuvuzela filters advertised but most fans just turned off the sound. FIFA, the governing body of football took the position that the vuvuzelas were part of the South African culture and supported the vuvuzelas and even sold vuvuzelas via their website.

In 2009 FIFA knew there was a problem but chose to ignore it and the global broadcasters did not object loud enough till it was too late. The sound of hives of vuvuzelas is not the expected sound of football and the irrational exuberance of a few hundred in the stadium was permitted while hundreds of thousands of television fans suffered the sonic assault.

I suggested two solutions to the sound designers of the 2010 World Cup in South Africa. One was sound supplementation, which was not taken seriously; the other to ban the vuvuzelas from inside the venues, which was not supported by FIFA management. Sound supplementation or replacement could be accomplished with vuvuzela-free crowd samples while ball kicks could be manually inserted from samplers in real time by operators or AI reproduction of clean ball kicks as was done in 2018.

Artificial Intelligence (AI) in Audio

Future sound production could easily see the use of AI to create listening augmentations and build a new audible world – pure audio augmented and virtual reality. It may be time to consider alternatives to sound design and spatial imaging and that's when it gets interesting with the capabilities of immersive sound. SALSA is a smart system that learns the field of play and optimizes the microphone coverage or inserts new samples in their place. See the section on artificial intelligence in audio in Chapter 3.

Front Soundfield Reinforcement (FSR)

Basic immersion can be accomplished by simple atmospheric enhancements and embellishment in the height channels; however, as sound designers move beyond fundamental immersive principles bold new and advanced alternative sound designs will be developed and embraced.

Alternative soundfield proposals must begin with soundfield deconstruction to understand advanced concepts of soundfield construction. Typically, immersive sound designers think in terms of two horizontal layers: one ear level and another above ear level. The Japanese broadcaster NHK added a layer below the ear level but still look at the soundfield as horizontal layers. This conceptualization may be from stereo and surround, which are considered to be horizontal, ear-level reproductions.

At the 2016 Tonmeister Conference in Cologne Germany, I presented the principles of a sound design that focuses the listener's attention forward and uses the height element to draw the listener's attention beyond the normal left and right peripherally of stereo and surround. With the principles of Front Soundfield Reinforcement (FSR) I am suggesting an additional correlated zone consisting of the front ear level and the front height channels. The soundfield in this front vertical slice/layer is the spatial relationships between the left, left height, right height and right speakers. The correlated use of these channels can create a cohesive reinforcement effect. For example, hearing the dialog did not seem to be a problem till surround sound. The stereo signal created a coherent centralized image of the sound from two directions but when we went to surround (and immersive) the dialog was relegated to the center channel with less frontal energy.

I am proposing using real and virtual sound images that are created from the correlation of four sound sources: left, right, left height and right height in front of the listener/viewer. Not only do these channels provide an FSR, but sounds can be steered and separated virtually anywhere in the front 2D vertical plane.

Positive soundfield principles use all speakers in the sound zone to reproduce the soundfield. The benefits are apparent when you use the height speakers in conjunction with the lower speakers to reinforce the soundfield with this intentional sound design.

FSRs are sound design principles that emphasize and reinforce the front perspective; just as with 4K television, where the picture stretches the boundaries of the picture, these principles stretch the sound proportional the screen from side to side and top to bottom.

It is easy for the ears to hear sounds from the vertical plane in front of and above the listener. My experience and listening leads me to believe that this soundfield construction matches the picture with a fuller, more detailed sound while keeping an immersive soundfield. Additionally, listening tests over up-firing soundbars seem to benefit from these principles as well.

Typical immersive sound design views the soundfield as two or three horizontal layers where the listener hears sounds at ear level as well as sound from above and below. The base layer is at ear level and is the horizontal layer of sound that also makes up the stereo, 5.1 and 7.1 surround foundation. Ideally the sound radiates to the listener close to ear level. Reflections should be considered. The height layer not only contributes to the sense of spatialization but also localization. The lower level was introduced with the 22.2 format. NHK 22.2 sound designs include spectral emphasis in the zones.

The sounds can be reproduced with actual and virtual speaker methods, however with each additional reproducer (speaker) there is a significant increase in resolution (detail) and at some point there are diminishing returns. Localization in the front vertical plane includes sounds from ear level and above the listener with a primary focus of the sound from the front left, center, right, left front height and right front height channels/speakers.

It is beneficial for sound designers to consider the parts that make up the whole 360-degree sound universe. If you divide/deconstruct the 360 universe into horizontal and vertical planes/zones you can examine the individual zones and determine how the zone should integrate and interact in the soundfield.

Defining the immersive soundfield as the various sub-components of the soundfield facilitates an additive approach to sound design. This means that by summing the soundfield sub-components, a holistic soundfield is derived while maintaining flexibility in controlling spatialization and localization.

The frontal zone consists of the left, right, left height and right height channels/speakers that are typically equal distant directly in front of the listener/viewer. The correlation of these four speakers creates a phantom image anywhere in the front rectangle and can reinforce the soundfield with clarity, detail, sonic quality and localization. The correlation between the speakers creates a directional forward vertical sound emitter working together as a concentrated source and power resonator.[4]

There are two lateral zones to the left and right of the listener. The left lateral zone consists of the left, left height, left surround and left surround height speakers. The right lateral zone consists of the right, right height, right surround and right height surround. The correlation of the left four speakers as well as the right four speakers creates a directional forward vertical sound emitter. As you move from the front soundfield to the sides, vertical correlation of the lateral soundfields improves side localization without using a real middle speaker.

The rear zone consists of the left surround, left height surround, right surround and right height surround speakers in the rear zone which are completely off axis of the listener.

The stratus zone is the four speakers above the listener – left front height, right front height, left rear height and right rear height. The height aspect of immersive sound has been an initial attraction to immersive sound production, but the question for some immersive sound productions is what do you use the height component for. The obvious sounds from above are airplanes, birds and weather but we are inundated with sounds that come from all around us – including from above. In daily life the sounds from above tend to be reflections of sounds and compounded by reflections turn to noise.

The Japanese advocate an additional zone below ear level. Although a completely valid approach there are too many reproduction issues at the time of this publication.

The 360-degree soundfield is a seamless fusion of direct sounds and complex acoustic reflections. There obviously is overlapping in the zones and care should be made to ensure the results are positive reinforcement and do not result in destructive phase problems.

Implementing V-Factor and Point of View

Define the sound design with a thorough evaluation of an event/sport for its vertical factor. What makes sense, what is expected, what is doable and what is audio effective?

The principles of FSR work with virtually any perspective and point of view. From extreme close-ups and compressed visual points of view with little room above the head of the athlete to sweeping wide shots looking up a mountain, cohesion in the frontal space contributes significantly to the believability of the scene.

Soundbar Reproduction

Front Soundfield Reinforcement will work well with up-firing soundbars. The height information will be directed in the up direction and, along with the reflections off the wall and ceiling, reinforce the frontal vertical sound image and constructively reinforce the entire soundfield.

Practical Applications

Specific use of the forward soundfields can be used to anchor the viewer to the front.

For example, Front Soundfield Reinforcement can improve speech intelligibility by hearing the voice in front of you imaged from all the front speakers – left, center, right, left height and right height. Clearly four speaker/channel reproduction is a different sonic representation than just hearing the vocal elements using the center speaker.

Remember, stereo adequately and clearly articulated the voice/vocal before a dedicated center speaker was introduced with 5.1 surround sound.

Sports with Exaggerated Horizontal and Vertical Movement

Often sports activities take place in the middle of a 16×9 high-definition screen where it is common to perceive lateral and vertical motion in the picture framing. This motion lends itself to sound embellishments to the sides of, as well as both below and above, the listener's ear level. For example, sports that are captured in a series of long panning shots with quick come/go, shots suitable for Doppler shift sound qualities. When the picture is wide 16 × 9 aspect ratio, lateral sweeping pans with sound enhances the production value and improves the sense of motion and speed.

With downhill skiing there are several shots where the skier is airborne, which permits interesting sound design possibilities to elevate some of the effects to reinforce the listener's attention to something that appears to be above the listener. The sound stretches the imagination up, down and laterally as the skier transverses the course. Any sport scheme can support minimum immersive sound design, which can be as simple as the addition of ambiance and atmosphere into the height speakers. This usually will create a sense of aural space for the picture, but advanced and interesting immersive sound design requires understanding the point of view of the picture coverage to determine the POV of the listener.

Basketball

Define your sound design by looking at the pictures. Basketball is a sport that has two distinct visual perspectives – a wide point of view of the game and the close point of view of the player. The pictures usually have a vertical aspect and my sound design is to emphasize this.

There is a correlation between spaciousness and the vertical separation between two dominant sound sources. For example, with basketball there is the sound of the net which visually appears above heads of players while the slap of the ball when dribbling appears to be at ground level. Mixing wide and close POVs are conducive to an uninterrupted immersive perspective without additional atmospheric enhancements. This picture lends itself to separating the crowd atmosphere above and the court bounces and squeaks in the lower quad. When you switch to the goal camera, the net microphones are in the height speakers and the close floor microphones are slightly below ear level and the upper atmosphere is reduced in the height channels. This sound design is not tedious or fatiguing to the viewer because the picture change and size justifies the changes in the soundfield.

Any sport that uses multiple different wide shots, such as downhill skiing, is perfect for FSR because of the wide camera perspective of the skier descending the slope. There is a tendency for the athlete to bump up and down, almost requiring the camera perspective to be wider. Additional sports-specific sound in the height speakers enhances the sports experience without being distracted.

Even sports like European football and rugby are conducive to head-to-toe reproduction of the sound to deliver realistic reproduction of the soundfield because the 3D soundfield is stationary during competition. Boxing is another sport that benefits from frontal reinforcement. The boxers are usually shot head to toe or with a loose wide shot. The impact of the glove is significantly enhanced with the frontal soundfield reinforcing the impact sounds. Since the 2008 Olympics I have injected glove sounds into the surround channels with positive feedback from peers and consumers. Moving the sound from behind to above the listener is a logical evolution in immersive sound design and frontal soundfield reinforcement.

In summary, a needs and solutions analysis defines and predicts what is doable and the probable outcome? Production and technical considerations must be accounted for in the probabilities of a successful production. Future-ready sound is open-ended technology developments and production possibilities and certainly next generation sound means something different to each consumer.

Future-ready sound is beyond proof of concepts, but the possibilities aren't fully understood by consumers and practitioners – yet.

Future-ready entertainment enhancements include a wide range of sound design that is controllable and scalable, but is more about the complete experience beyond the picture.

Notes

1 Dennis Baxter. 2020. "Faking It: Sound-Starved Stadiums." TV Technology. November 5, 2020. www. tvtechnology.com/features/faking-it-sound-starved-stadiums.
2 Baudrillard, Jean. 1981. *Simulacra and Simulation*. Ann Arbor: The University of Michigan Press.
3 "The Sound of Sports." 99% Invisible. https://99percentinvisible.org/episode/the-sound-of-sports.
4 Dennis Baxter, "Spatial Sound" (presentation, Tonmeistertagung 2016, Cologne, Germany, November 17–20, 2016).

3 Mixing and Spatializing Future Generation Audio Content

Effective immersive sound design includes the fundamental application of localization, spatialization and advance audio production practices (AAAP). This chapter will present principles and methods for creating three-dimensional soundfields using audio objects, spatializer tools, 3D panning, advance simulation techniques and higher order ambisonics for localization and spatialization.

Channel, object-based and ambisonic methods of audio production are all capable of producing immersive sound with different levels of accuracy and believability. All methods and theories of immersive sound production require the ability to localize a sound element in a dimensional sound-sphere, therefore 3D localization is one of the fundamental principles of spatialization. Spatialization is the process of enhancing the aural architecture of a soundscape not only with the hardware, software and sound shaping tools, but also applying the research and intellectual resources for results beyond those predicted.

Immersive sound is created from sound elements that are mono, stereo and/or multidimensional and in the process of creating immersive sound these core sound elements are fused with advance spatialization enrichments resulting in an enhanced and expanded dimensional soundfield.

Advance Audio Production Techniques/Practices (AAPTP)

Advance Audio Production Techniques/Practices (AAPTP) is an umbrella of Advance Production Tools and Methods that incorporates spatial imaging and modeling that coalesce the spatial sound image. These production practices function together in line with the fundamental mixing operations of processing, blending, combing and routing the audio signals.

Benefits of localization are beyond the actual placement of sound objects. For example, spatial organization and separation can significantly improve sonic clarity by unmasking audio elements. The ear can easily differentiate sounds that are only separated by a few degrees from each other. Unmasking sound elements allows them to be separated and easily heard without increasing volume levels.

The use of panning, specifically elevation, helps tie the lower layer of sound to the upper layer of sound by adding components of each sound layer into the adjacent soundfield. Slight correlation between the zones – the horizontal zone – left-right, front-back and up-down create phantom images and has been proven to positively reinforce the soundfields.

> **Production Note**
>
> It is desirable to keep a stable sonic foundation of ambiance and atmosphere that should change little compared to the specific sounds embellishments.

DOI: 10.4324/9781003052876-3

Precise Localization: Perceived Localization

With sports production precise localization of sound elements such as atmosphere and ambiance is not essential to realistically create immersive soundfields. Typically, ambiance and atmosphere productions benefit little from spatial enhancements while close, detailed sounds do. Spatial perspective can be created (and adjusted) by positioning the microphone(s) to capture a good blend of direct and reflective sound.

The Mixing Surface (Platform): Real or Virtual

As of 2022, the current array of engineering and production options provides the sound designer with the ability to produce immersive, interactive and personalized sound in real time on a large-scale software-driven mixing desk as well as on computer-based work stations. The mixing surface is the heart for all signal management, routing and the platform to bring an open architecture into play for additional resources such as further process sounds through plug-ins (see Chapter 4).

Mixing, sound design and audio management for post-production and live real-time mode is considerably different. Complex soundscapes with movement and high dimensionality are easily accomplished in post-production where live soundscapes that convey motion and speed are a combination of effective microphone placement and processing. For example, a well-placed microphone on a ski slope along with some processing with doppler effect can easily convey motion and speed in real time.

Spatialization, localization and advance audio production practices require a platform to operate. In our digital world fundamental sound management is usually performed on a computer platform where basic functions such as balancing, combining, processing and routing sound signals is usually executed in a computer/virtual domain with user interfaces to replicate a familiar workflow. For example, the tactile experience of touching a fader or turning a knob can be real or virtual on a screen depending on user preference, space and budget. For a live production where instantaneous reactions may be required, it is not conducive to drilling through layers of virtual faders and menus to make an adjustment.

Regardless of whether the mixing platform is real or virtual, sound signal management and mixing platforms usually are built on a hierarchy of channels that are assigned to group and busses which are further assigned to output channels. It is the ability to build several multichannel busses and groups that effectively builds layers which are necessary for complex immersive soundfields. For example, your control surface would integrate a high order ambisonic microphone into a dimensional mix through its own unprocessed bus where mono and stereo microphones would be processed and assigned to a separate immersive sound bus. Multi busses allows for accuracy in level and timing (phase), as well as an effective method to build and modify soundfields simply by combining complex soundfields.

Where Do We Start?

The mixing/control surface will perform the functions of equalization, dynamics modification and localization. Later in this chapter we will look at the use of equalization and modeling as a spatialization technique. A fundamental principle of spatialization is separation. When listening over two or more loudspeakers, the level difference between the channels is perceived as the direction of the source.

As audio professionals, we are familiar with left/right and front/back horizontal panning. For example, when a mono signal is being panned to two or more channels you are adjusting the ratio of the amplitude (volume) between the two channels. Panning the signal to the left increases the amplitude of the signal in the left channel while the amplitude of the signal over the right channel is decreased. Panning the signal to the right increases the amplitude of the signal to the right and decreases the signal to the left. When leaving the signal (amplitude) equally to the left and right or the center channels, the listener perceives a virtual sound source between the speakers. This panning method is known as Pair Wise Amplitude Panning and was discussed by Blumlein in 1931.[1]

Spatialization can be viewed as the order or degree of separation. The first order of separation is from mono to stereo where sounds can be clearly segregated into two channels. The second order of separation is sound production moving from stereo to surround. But even with stereo and surround sound the audio is only reproduced on the horizontal plane although in front, to the sides and behind the listener. It has also been recognized that the practice of adding a few degrees of separation vertically (height) gives the third order of separation among sound components, significantly expanding the apparent soundfield.

Horizontal surround panning is capable of placing a sound object/element anywhere in the 360-degree soundspace around the listener with accurate localization in the horizontal plane. Horizontal panning can image a sound anywhere around the listener including directly to a speaker or between adjacent speakers creating a phantom image between two speakers.

Panoramic separation, localizing, balancing, combining, processing and routing the audio signals are central to audio management and production and are generally accomplished within the framework of a mixing desk, either real or virtual. 3D Panning is the most fundamental spatializer tool that gives the sound designer/mixer the means with which to accurately position the apparent fixed location of a sound anywhere in the soundsphere. These three-dimensional panner/positioner tools are usually applied to audio elements to position them in any 3D coordinate/position in the acoustic soundscape – around, above and below the listener. 3D panning is the fundamental spatialization tool that can locate sound horizontally in the XY axis – left, right and front and back as well as the vertical Z axis – up and down, full azimuth and elevation control for precision.

In the digital world, DSP digital sound processing makes sound location computations using Cartesian coordinates (azimuth and elevation) that are used to mathematically specify the location of a point in three-dimension space. Any location on the horizontal axis can be designated as an azimuth location. Since the concept of an immersive soundfield means spherical reproduction, in an immersive world the azimuth control rotates the sound element/object 360 degrees around the listener.

But the fact is that for truly dimensional positioning, a height component is also necessary. The vertical aspect of panning tools moves the audio up and down on the perpendicular axis. Vertical panning is known as elevation.

Vertical panning/elevation is effective because additional height dimension contributes to the detail and localization of sound elements/objects in the soundfield. As you add elevation to a channel/audio element, the adjacent channels, above and to the side of the sound, contribute to the amplitude and positioning of the soundfield.

Horizontal and vertical panning moves the sound around and above the listener and is effective for sonic separation of sound elements. Panoramic separation (also known as panning) is moving or rotating the sound element anywhere around the listener including directly to a speaker or between adjacent speakers creating a phantom image between two or more speakers.

By definition, a 3D speaker setup means that not all the loudspeakers are in the same plane – some speaker(s) have been elevated. Spatialized panning in a three-dimensional loudspeaker setup can be accomplished by various rendering technologies. Two that are well known are vector based audio panning (VBAP) as well as various ambisonic rendering methods. VBAP[2] achieves a location coordinate by using up to three speakers forming a triangle from the listener's point of view formulating triplet-wise panning between three loudspeakers.

The listener will perceive a virtual source inside that triangle depending on the ratio of the loudspeaker's amplitudes. Virtual sources, known as phantom images, are easily created by applying sounds to all three speakers or some subset of the speakers when more speakers are present. Multiple virtual sources can be applied to each triplet.

Triplet-wise panning can only be applied to a maximum of three speakers (channels) at a time. When the number of loudspeakers is greater than three, triplet-wise triangles are recalculated for the number of speakers so the triangles are not overlapping.

Ambisonics is a different approach to dissecting the soundfield and a different approach to sound design. With Ambisonic, the sound is applied to every loudspeaker/output buss at the same time with the panning device controlling the ratio of the sound between the speakers/channels. Ambisonics panning works well because with ambisonic production the goal is to recreate as much of the soundfield as possible and a sounds localization is virtualized within that soundfield no matter how many or how few loudspeakers are available to the listener. When you increase the number of soundfield coefficients (speakers/microphones), you increase the audio resolution of the soundfield resulting in greater detail, sonic clarity, localization and phantom imaging.

Production Note

If you consider the front soundfield – left, center, right, left height and right height and using this forward soundspace to reinforce the positive elements of your sound production then your sound production possibilities are expanded. For example, with ambisonics there is the possibility for improving speech intelligibility by rendering the speech channel into the front soundfield and not just localized in the center channel.

Benefits of localization are beyond the actual placement of sound objects. For example, spatial organization and separation can significantly improve sonic clarity by unmasking audio elements. The ear can easily differentiate sounds that are separated a few degrees from each other. Separating sounds can give better clarity and detail to the soundfield by unmasking sound that loses definition in the mix. Unmasking sounds improves the precision of all sound elements by contributing to the definition, clarity and positioning of all elements in the soundfield. Unmasking sound elements allows them to be separated and easily heard without increasing volume levels.

Additionally, the use of panning, specifically elevation, helps tie the lower layer of sound to the upper layer of sound by adding components of the sound to the adjacent soundfield. Slight correlation between the zones – the horizontal zone – left-right, front-back and up-down create phantom images and has proven to positively reinforce the soundfields.

Live Immersive Sound Production

Constructing an immersive mix on mixing console with 8 buss architecture has been accomplished since 2018. The realities are that 3D panning requires a minimum of eight channels with the lowest resolution of two channels of height channels – 5.1.2. Even though it seems that an 8 buss output architecture is a compromise, creating concentric 8 channel busses will satisfy many of the immersive needs for first generation immersive sound for sports. National Hot Rod Association has produced 5.1.4 immersive sound since 2018 using a Calrec

8 channel bus configuration using overlaying concentric 8 channel bus configurations and pans between the 5.1 layers (see Chapter 10, Case study 10.18).

Either production method, quasi-correlated layers or true correlation between the layers will ultimately results in an immersive (5.1.2, 5.1.4 or 9.1) audio format. It is expected that the mixing console will do the heavy lifting for spatialization control in the live environment. Live broadcast and particularly sports are operator driven which requires a different workflow than studio driven workflows such as episodic and film content.

Most live and post-production mixing platforms for immersive sound have developed to a minimum of 12 busses with X, Y and Z three-dimensional correlated panning. Twelve channels could deliver 7.1.4, a higher potential resolution of sound.

This chapter will present competing console manufacturers but emphasizes the wide range of desirable features that facilitate high channel density sound production.

Solid State Logic

Solid State Logic was firmly entrenched in the high-end broadcast industry early and has continued with the System T that support immersive audio infrastructure including Dolby Atmos, MPEG-H, ATSC 3.0, as well as other 3D positioning. SSL System T natively supports any channel or bus format up to 7.1.4. The System T monitor section shares many functions that will be familiar to SSL users, but have been expended to include support for multiple 12 channel outputs, formatted up to 7.1.4 as well as 7.1.4 to stereo fold down.

Monitor section key features: 49 × 12 Channel Monitor Inputs Up to 12 channels to support 7.1, 5.1.2, 5.1.4, 7.1.2, 7.1.4 and 4.0.4 monitor formats. Four sets of monitor speakers: two sets of 12 channel Monitor outputs plus two sets of stereo nearfield outputs. Additional Stereo PFL Monitor Output Dual 12 channel monitor insert points. 2 × 24 external sources selectors + primary input. External source selector outputs available as routable.

All EQ's, dynamics and effect rack processors operate in native formats up to 7.1.4 with individual level and delay compensation for both main monitor outputs. Individual-level compensation for both all outputs and polarity invert on all speaker outputs. Simultaneous monitoring of two independent program feeds as well as Active Solo Insert and Alt Stereo Monitor override.

Systems T's immersive and 3D toolkit includes Ambisonic transcoding tools and Binaural encoding tools. The onboard ambisonic 360 transcoder effect unit takes A or B format inputs and renders to a wide range of multichannel outputs from 4.0 to 7.1.4. Included A-B conversion for the AMBEO VR Mic directly outputs the processing-ready B-format, for rendering to multichannel, allowing our clients to use the Ambeo VR Mic to add ambient sound quickly and simply for live immersive productions. The transcoder additionally provides decoding for a live VR mix or B format archive recording.

Broadcast engineers will be able to record and process immersive content with far greater ease, and without the need for any additional physical equipment. The pan program can automatically adjust width parameters to achieve the desired pan position.

All System T paths (channels and busses) can be configured in formats up to 7.1.4. When a channel or stem path feeds a bus, the panning parameters for the largest bus being fed become available. All panning is observer to all bus formats, so a channel routed to both a 5.1.4 bus and a stereo bus will pan correctly to both. When bus routing involves downmixing, two fold-down options are available per path. Adjusting a path's front/rear pan will automatically adjust the depth parameter to achieve the desired front/rear pan position.

Two panning modes per path are possible, XYZ and ThetaPan. Both panning modes allowing point source signals to be positioned in a 3D space, but also incoming or sub mixed multichannel signals to be manipulated and then positioned.

Width and depth automatically adjust – as required when Pan controls are adjusted.

Examples we have seen used include an ORTF 3D mic or Hamazaki square arrangement in venues and stadiums, brought into appropriate format channels, perhaps reducing width, depth and height (where applicable). This results in being able to use pan, front/rear and bottom/top controls to position the resulting signal.

For XYZ panning the traditional XY (e.g. 5.1) Pan LR, width, divergence and depth controls are expanded to include btm/top and height controls in the Z domain.

Pan moves a mono signal left and right. For signals with left and right components pan is progressively proportional to the width parameter.

Width is available when both the channel and bus have left and right components. The control has a range of 100 to -100 percent, reducing the width of a stereo signal to mono at 0 percent. As the width becomes negative the image is widened back to full stereo but inverted. For busses with a center component, as the width is reduced the mono component is redirected to the center channel. A reversed image is indicated by the arrows in the display turning red. If Width is 100 or -100 percent the L/R Pan has no function. As width is reduced, the L/R Pan acts as an image offset, by shifting the narrowed stereo image between the edges of the soundfield. If the width is 0 (mono), the L/R Pan acts as conventional mono pan pot.

Figure 3.1 XYZ panning mode and ThetaPan panning mode

Divergence is available when the bus has left, right and center components. As divergence is increased, the center channel is progressively added to the left and right busses, at the same time reducing the center channel gain. 100 percent divergence provides a true phantom stereo image. When the channel has left and right components, divergence interacts with width to determine whether the narrowed stereo image has a phantom or discrete center component.

Front/Rear (Frt/Rr) is available when the channel or stem does not have a rear component and is routed to a bus which does. It acts as a surround panner moving the signal forwards and backwards in the soundfield.

Depth is available when both the channel or stem and bus have rear surround components. Depth acts as a differential balance control for the rear channels in relation to the front channels. As the depth is reduced, the rear channels are proportionally added to the front channels while being attenuated in the rear channels. As with width, the Frt/Rr pan progressively acts on the reduced soundfield.

F/R divergence is for the side speakers, i.e. valid 7.1, 7.1.2 and 7.1.4 formats only, and determines the percent of signal sent to the side speakers. A 100 percent divergence is equivalent to 5.1, i.e. no signal is sent to the side speakers when the front/rear position is F50/50R, the signal is distributed 50/50 between the front and rear speakers. 0 percent divergence means that the side speakers receive the full signal at F50/50R position.

Low frequency gain (LF gain) is available when the bus has an LFE component; it provides a level control for the channel contribution to the LFE bus.

Bottom/top is available when the bus has height components, moving the signal up or down in the 3D soundfield provided the channel or stem does not have a height component.

Height is used on channels and stems with a height component that is routed to busses with height components, i.e. the 5.1.2, 5.1.4, 7.1.2, 7.1.4 and 4.0.4 formats. It is equivalent to the depth parameter that is used to collapse a 5.1 channel in the horizontal plane but height acts in the vertical plane. As with depth, which collapses the sound field from the rear, the height initial value is 100 percent and collapses the sound field from the top as it is reduced, eliminating the height component entirely at 0 percent. As with width and depth, Btm/Top becomes progressively active as the height is reduced.

Low frequency only (LF only) is available on any path with a format that includes an LF channel when routed to a bus with a format that includes an LF channel. Activating LF Only removes all contributions to the other busses leaving the path feeding the LF bus only via the LF gain parameter. For stereo and above formats the path legs are summed prior to feeding the LF gain control.

Figure 3.2 Ambisonic signal flow in System T mixing console

For ThetaPan the controls switch to rotational controls that elegantly spread a mono point source across adjacent speakers without requiring complex multiple divergence controls. Additionally, it enables signals to be passed throughout the boundaries of the soundfield in a circular motion.

- Angle rotates any signal between −180 and +180 degrees.
- Spread is available for mono point sources; it spreads the signal across adjacent speakers around a 360-degree circle.
- Width is available when the channel has a left and right component. Starting at 100 percent for 1:1 mapping, width can be reduced to focus signal, or increased to move signal away from the focused angle.
- Bottom/top and height controls in Theta pan have similar functionality to XYZ panning, manipulating the signal vertically within a cylindrical representation.

Input formats from 4.0 to 7.1.4 feed the binaural encoder to monitor Dolby Atmos and MPEG-H and AC-4 in discrete speakers or by headphones. The Binaural 3D Encoder effect unit expands System T's immersive and 3D audio toolkit. With input formats from 4.0 to 7.1.4, the binaural encoder provides the ability to monitor immersive beds for Dolby Atmos or MPEG-H via headphones where additional speakers are not available or provide a stereo binaural feed for streaming purposes. The binaural 3D encoder currently uses the SADIE II HRTF dataset.

MS decoder available in the input sections of a stereo path (optional) and MS control provides a differential gain between the mid signal on the left input and the side signal on the right input.

The 360 transcoder includes coefficients to directly support the first order format and directly supports the Sennheiser Ambio. The input stage provides for orientation of the microphone, end-fire and invert to accurately position the microphone for decoding. Rotate/Zoom and tilt controls/Center channel control/Transcoder Format Width/Phase control in downmix/Change positions of virtual speakers

Range of multichannel outputs formats. The 360 Transcoder can be used in real time with A Format to provide an ambient bed for live production and can use B Format from previously converted recordings. Both FuMa and AmbiX formats can be decoded.

Formatted busses (e.g. Stem subgroups) can be mixed (to master busses) without timing issues.[3]

Calrec/DiGiCo

Calrec has been a stalwart in the broadcast industry for decades and has managed to keep ahead of the most dynamic changes ever in the audio industry. Calrec always had a reputation for robust design and was a fixture with BBC sports and the dominant mixing desk in their OB fleet.

Calrec's Bluefin3 IP engine has increased input channel leg capacity but requires an audio processing and routing engine for immersive sound. ImPulse is the immersive software integral in Calrec's 3D audio processing and routing engine. ImPulse feature set includes height legs, path widths, 3D pan controls and more flexible downmixing. There is no restriction to any ImPulse user; the only restriction is the amount of DSP in different pack sizes − Calrec offer a range of packs which have a range of DSP amounts. ImPulse supports mono, stereo, 5.1, 5.1.2, 7.1, 7.1.2 and 7.1.4 paths on input channels, groups (+0.0.2 and 0.0.4 height busses), main paths along with associated monitoring and metering. All path formats can coexist − Impulse has flexible downmix and surround panning between formats. ImPulse is compliant with SMPTE 2110 connectivity and can be added to existing Apollo and Artemis control surfaces to provide an upgrade path for existing Calrec customers as they transfer to IP infrastructures. In that respect it can be used on Apollo and Artemis consoles, but not Summa, Brio or Type R.

Immersive mixing is a method to place a specific source within an immersive mix, requiring a 3D pan allowing placement of an audio source within the 5.1 or 7.1 bed but also with a height element.

This can be achieved with a two 5.1 mix busses sound design, but ImPulse also has 0.0.2 and 0.0.4 height busses to add height to a standard 5.1 input. This is much more efficient than using 2 × 5.1 channels to accommodate immersive sources which effectively uses 2 × mono channels.

It is useful to have not only a touch interface allowing the operator to pan a source anywhere within the image but also individual controls to pan between the 5.1 and height beds allowing individual sources to be easily placed either only within the 5.1 bed or only within the height bed.

Just as in 5.1 panning the option to have a divergence control to spread a specific source within the sound image is a useful tool.

SPILL – on a Calrec console when an immersive channel (or 5.1 channel) is patched to a channel on the console it is controlled on a single fader; the overall level of that signal can be controlled by that single fader, so moving the fader up will increase the level of all components that make up the immersive signal, and altering EQ or dynamics will apply that same processing to all components. Spill faders provide control over any of the individual components that make up that immersive or 5.1 channel. In other words, the immersive signal can "spill out" onto the spill faders. This allows individual-level changes to be made and processing applied to each component (i.e. L, R, C, Rl, Rr, LFE, plus the height components), rather than the overall signal. Spill can be applied to any immersive fader it is assigned to.

Immersive spill – ImPulse makes very efficient use of surface controls for accessing the individual elements of immersive paths. Only one fader is required on the surface for global manipulation and application of signal processing to the complete collection of paths that make up that signal. Up to 12 mono paths for a 7.1.4 signal are required and controlled simultaneously by a single fader. Under normal circumstances this provides an extremely quick way to alter the level or assign processing to all component paths of a given signal. For times when more flexibility or greater control of individual component signals is required, the signal can be expanded onto the spill faders. Bus widths can be configured from mono up to 7.1.4 wide, so a 7.1.4 wide immersive bus uses 12 mono DSP resources.

Immersive spill legs – if an immersive master is part of a VCA group then the VCA primary and secondary master levels, cut settings and APFL settings affect all of its spill legs. It is not possible for the spill faders themselves to be masters or slaves of a VCA group.

Downmixes (more on downmixing below) – with the addition of immersive busses a further level of downmixing is made available. For bus widths above 5.1, e.g. 7.1 and above, a 5.1 downmix output is provided. Each 5.1 bus is provided with a stereo LoRo down mix which maximizes mono compatibility.

Controls – depending on the width of the current path, and the width of the destination bus, different pan controls will be made available on the surface. For example, when sending a mono path to a mono bus, there will be no pan controls available. When sending a mono path to an immersive bus, panning controls will be presented that allow control over mono placement in a surround/immersive field. The complete range of controls is described here:

- Front pan allows positioning of the signal in the L and R speakers.
- Center only (sets the signal to appear only in the center speaker. It effectively overrides all left and right pan positions).

- Front L-C-R (pushing the FRONT L-C-R button switches the FRONT PAN control between L-R panning and L-C-R panning).
- Front divergence (with FRONT L-C-R switched in, the spread of the signal can range from fully converged in the C speaker, through equal level in L, C and R, right the way to full divergence with no level in the C speaker and full level in the L and R speakers. The button next to the divergence rotary control switches the divergence position in or out).
- Front-rear pan (FRONT-REAR PAN varies the position of the signal between the front and rear speakers. The L and R position in the front and rear speakers is independent and can be controller separately with the FRONT PAN and REAR PAN controls. The signal is moved from the front pan position through to the rear pan position). Rear pan (allows the left to right position of the signal to be set in the rear speakers. This is independent from the front pan position and can be switched in or out).
- The LFE control varies the level sent to the LFE speaker. Non LFE (the level sent to all channels other than the LFE channel and can be varied with the NON LFE control. When this control is switched out and the signal is panned to one or more channels other than the LFE, the signal is sent at full level).
- 3D controls – in addition to the above controls, the immersive paths add the requirement for height panning so that a source can be positioned anywhere in a 3D X-Y-Z space. Two controls have been added to the previous panning controls – height pan and 3D view. Height pan: Height pan allows positioning of the signal vertically in the Z plane from the 2D surround field of 9.0 or 7.1 speaker arrays into a 3D immersive field with additional height speakers either in a two-speaker arrangement.

This button toggles the view between the normal send and route display on the control surface into a 3D room view image to show where the sound source is with respect to the left-right X axis, front-rear Y axis and height (up-down) Z axis. When pressed it displays the image shown above.

The eight corners of the room are labeled to show the orientation of the view. The floor of the room represents the X and Y axes that would be applied to a surround space with an additional center reference for the front LCR arrangement and Ls/Rs which represents the back of the surround space. This is also displayed in a 2D display labeled Surr (X-Y), to the right of the room view.

The top four corners of the room represent Left top front (Ltf)/Right top front (Rtf) at the front and Left top back (Ltb)/Right top back (Rtb) placed at ceiling height. The red ball represents the position of the source within the 3D immersive space and the lines extending from the sides of the ball show the amount of front divergence being applied. When height is added to the red ball position, the ball will move vertically to show this, which is also shown in a 1D display labeled height (Z) to the right of the room view. Above this is also shown the stereo L/R position the source would appear in a stereo field. Note the stereo, surround and height displays are also shown on the sends/routes view.

With the 3D view button active, the 3D view image appears for a short period and then returns to the normal send & route view to allow routing functions to continue; however, if any PAN control that is switched ON has its pan control position altered, the 3D view returns until any control changes stop, at which point, after a slight delay, the display returns to the normal send & route page.

Joystick Panning

In addition to using the pan controls in Send-Route, Strips and Wilds mode (if assigned), panning can be achieved using the joystick panel, if fitted. "Strips mode" is like an analogue

Figure 3.3 Calrec 3D view

console with control parameters lined up above the fader strip, and "wilds mode" is an assignable system where all the control parameters apply to whatever fader is selected.

When the system is in Show height mode for immersive mixing, the joystick function changes from left–right pan on the X axis control and front–rear pan on the Y axis control to left–right pan on the X axis control and height pan on the Y axis control.

Downmixes

ImPulse has comprehensive automatic downmix facilities to make downmixing as simple as possible while retaining complex control where necessary. Downmixes are applied to the metering and monitoring systems, including APFL and all relevant paths in the system. When an immersive path, metering or monitoring signal is routed to a mono, stereo or surround destination, a downmix must be applied to take care of the increase in level that will occur due to the summing of immersive components, and also to shape the sound in the desired way.

ImPulse allows a number of default downmix configurations to be set up. One of these downmix configurations can be selected for use in a show. All downmixes performed within that show will follow the default settings selected, unless changes are made to individual paths or busses by the user. The downmix configuration is applied to downmix faders automatically for all immersive paths and changes to the downmix faders for an individual path can be made and applied only to that path.

Should an immersive path or bus be sent to a bus pre-fader, it may be necessary to have the spill fader levels applied to make the downmix the same as the post-fader downmix. This is because the post-fader downmix is processed after the spill faders, and so adjustments to individual legs on the spill faders will have an effect on the resultant downmix.

In some circumstances, it may be preferable not to downmix elements of a surround path to a certain bus, for example if a presenter is being fed an immersive source into their mono earpiece, it may be beneficial to omit the rear and height channels to enhance the clarity of any dialogue in the front. In this case, electing the front L/R (or center) element of the immersive path from the spill panel makes the front L/R only the currently assigned path and allows it to be routed on its own to the required destination.

When an immersive path is routed to a narrower bus from its master, rather than from a spill leg, a downmix is automatically applied. Immersive stereo downmixes maintain stereo separation by mixing front L, rear L, C, top L front & rear to create a Lo – "Left Overall" channel, and front R, rear R, C, top R front & rear to create a Ro – "Right Overall". Default downmixes do not include the LFE channel, but this can be added if required. The levels that the surround/immersive elements mix together to form a downmix are pre-defined and applied console wide, but operators can select from five other pre-defined downmix sets. Changes take effect immediately and are applied to all user memories within the current show. As well as downmixing for internal system routing, surround or immersive main outputs, all surround or immersive channel/group direct outputs have downmixed versions available for patching to output transmitters.

ImPulse also has a 'Reset all values to the Calrec defaults' button which overwrites all current downmix values with default values to provide an easy way to clear any unwanted edits and start creating a new downmix setup from a known default set of values.

The first ever live broadcast in Dolby ATMOS was NBC's Opening Ceremonies at the Rio Olympic Games in 2016. This was done on a Calrec Summa console. As an aside, Jayson Polansky did not use Dolby's Bass Management system for loudness and the Summa direct outs were uses to feed back into the console with adjustments made.[4,5]

Lawo

Lawo has supported immersive audio for many years including its development of a 22.2-channel audio system for NHK's Super Hi-Vision production as well as providing the mixing platform for the immersive audio production of the Men's World Cup finals in 2018 from Russia.

Lawo's AUHD Core Audio Engine offers native support for existing 3D/ immersive audio formats, such as 5.1.4, 7.1.4 up to 22.2 with integrated multi-bus 3D/immersive mixing in a variety of configurations. A__UHD Core supports Next Generation Audio formats such as Dolby Atmos and MPEG-H and provides comprehensive authentication for metadata which is necessary in the 3D/Next Generation Audio production process.

The Lawo mc²96 and mc²56 consoles manage up to 1,024 DSP channels and 256 summing busses. The consoles include Selective Recall, Oversnaps (relative trim-sets), and enhanced signal management functions for large productions including Swap and Relocate. Comprehensive audio-follows-video functionality for fast-paced, automated audio/video crossfades. Audio-follows-video is based on allocating each camera tally to an event that can be assigned to the desired number of audio channels. Up to 128 events can be defined, and each channel provides an envelope with adjustable rise-time, on-time, hold-time, max-time and fall-time parameters.

Integration of WAVES MultiRack SoundGrid multichannel effects processing in the mixing desk. The system gives operators access to Waves' comprehensive plug-in library, allowing them to conveniently control reverbs, multi-tap delays, graphic equalizers and multiband compressors directly via the console's keyboard and touchscreen. All plug-in settings can be easily stored and recalled with the console's snapshot and production files.

KICK

Lawo KICK software uses tracking data such as Chyron Hego's camera-based TRACAB tracking system, which generates the match data and feeds the required information to KICK's automation system: Lawo's KICK software acts as the interface between the incoming tracking data and the digital mixing console that dynamically sets audio channel levels.

The camera-based technology generates real-time data that tracks the exact position of the players, referees and the ball, and uses the existing microphones, their location on the edge of the pitch and their directional characteristics to determine which microphone is best placed to capture what is happening on the field. The system uses live tracking data of the players to set the correct input levels for the microphones, and repeatedly and automatically sends the necessary commands to the mixing console to open the appropriate microphone channels. KICK's unique technology provides reliably consistent, fully automated, high-quality, close-ball audio mixes for sports such as football, rugby and American football.

The system delivers transparent sound pickup with its excellent 'kick-to-noise' ratio which significantly reduces ambient crowd noise. It also ensures a consistent audio level without noticeable fades for seamless inclusion in a broadcast mix that is 100 percent repeatable from match to match and highly accurate. KICK's intuitive graphical user interface allows easy adjustment of all its parameters, including the placement of microphones, their polar patterns and microphone prioritization.

Figure 3.4 Lawo KICK screen

KICK is available as a software-only solution for productions using Lawo mc^2 mixing consoles, as well as a hardware bundled solution for productions using consoles from other manufacturers. It requires a standard host PC for the host application, and an HTML5-capable browser for the user interface. An optional hardware processing engine is available for interfacing to third-party consoles via RAVENNA/AES67.

Parallel compression is now available for all Lawo consoles, and can be applied in every channel, group, aux and sum of the mc^{2}96. Parallel compression (also known as New York compression) is a dynamic range compression technique achieved by blending a dry signal with a compressed version of the same signal. Rather than bringing down the highest peaks for the purpose of dynamic range reduction, it reduces the dynamic range by bringing up the softer sounds, which results in added audible detail.

Additionally Lawo consoles have direct integration of Neumann DMI-8® digital microphones into the console, which allows adjustment of gain, pre-attenuation, polar patterns, low cut filter settings and others, as well as direct integration into the console of the Waves Soundgrid®. The system gives operators access to Waves' comprehensive plug-in library, allowing them to conveniently control reverbs, multi-tap delays, graphic equalizers and multiband compressors directly via the console's keyboard and touchscreen.[6]

Production Note

BT Sports and Football

With any sport (event) minimum immersive sound production can be as simple as injecting additional ambiance and atmosphere into the height speakers such as what has been done at the 2021 Tokyo Olympics and at World Cup Football. Some sound above the viewer/listener will usually create a sense of aural space for the 2D picture (even at 4K resolution). To me football is the definitive example of the use of overhead atmospheric enhancements for spatialization. Additionally, overhead atmospheric enhancements does not require any localization and can be accomplished on most 5.1 mixing desks.

Note: 3D panning is helpful for precise localization, but bottom line – convincing immersive sound for most sports does not require precise localization. Additionally, there is no doubt that these field sports have no "sports-relevant" sound in the vertical axis (above the listener) only ambiance and atmosphere and artificial embellishments are probably not appropriate (see Chapter 10 for a complete case study of BT Sports).

Artificial Intelligence in Audio: Are We There Yet? AI or A1?

Artificial intelligence (AI) is used at Wimbledon. The computer watches the broadcast feeds and identifies interesting indicators by applying a variety of metrics. The metrics guides the computer in learning how to recognize significant points of interest and what makes a good highlight or replay. Interestingly, sound is a leading and reliable indicator. For example, pandemonium in the crowd after a long quiet pause is a good indicator of a memorable highlight moment. One of my logic metrics would also include the duration of the crowd burst as well as the amplitude, threshold, attack and sustain of the sounds during the interesting moment.

Artificial intelligence should be good at analyzing repetitive patterns and picking the best choice(s) from a set of indictors. For example, a very loud, sudden burst of crowd (sharp attack) with a long sustain is probably a good indication of a goal. The voice inflection of the crowd – sustained screaming as opposed to a sigh of dismay that dies out quickly – is another valuable and identifiable metric. From these simple learning indicators the computer, within a dozen repetitions or even 100 times, will be able to accurately predict a good highlight moment.

I would argue that in 2018 we were close to something. Lawo developed a mixing system that takes data of the ball position and translates that into capturing the best possible sound from the best possible microphone or combination of microphones plus determines what level to mix and blend them together. Tracking the ball is done optically and in a sport like football, the focus of the game is the ball – basically you tell the computer to follow the ball.

Arguably 2018 was the best sounding World Cup I have ever heard, and I would even say that it was the best sounding football event ever. Praise goes to HBS Christian Gobbel, Felix Krückels and the Lawo team for implementing a true paradigm shift in the world of sound for broadcasting but I think Philipp Lawo is on to something else.[7]

Now let's take a look into the possibilities of AI for sports coverage. Artificial intelligence comes into play when a computer analyses the switching patterns of a sample of directing styles and compares the director's commands to the position of the ball within the field of view of the broadcast cameras. The computer archives the director's selection for future learning. Within a short period of time, patterns will be detected, examined and programmed into event cycles to take over the direction of the cameras. A basic "follow the ball" pattern is learned, however it would seem possible that you can modify the production by blending and altering production styles. I once worked with a director who had a rhythm and repetition to his cutting style and literally repeated a dozen or so patterns over the course of the game.

I can clearly envision the day when bots and droid-computers capture, direct and produce live sporting events with little human intervention. Let's follow the flow. Camera robotics have been around for a while and there is no reason the cameras and audio cannot follow the electronic commands of a computer that is following play action. Imagine this possible scenario – the computer is calculating that after a goal kick, 7 out of 10 directors would cut to a wide shot while optical position tracking is continually sending the directoid mapping data of the field of play. The directoid directs camera x, y and z to be following the ball while simultaneously directing camera A and B to track the coaches. Additionally, camera A and camera B are capturing the audio from the coaches and sending the information to the directoid that is learning the patterns of the coaches and learning when to cut to the coach. The directoid has a library of possibilities for each ball position and makes comparisons.

Real-time action coverage could include speech interpretation from live commentators. The computer can learn cues from the commentators and filter this information to incorporate action cues that break away for relevant action away from the ball.

Alternatively, instead of digitizing speech for the computer to use the data, the computer can ingest all the data and artificially create the commentary track. Speech synthesis has been around a while and once you have optical tracking it becomes conceivable that you can create droid commentators that interpret the play-by-play action and sound re-synthesis to complete the entire experience – alternative reality.[8]

Spatial Automated Live Sports Audio (SALSA)

Dr. Ben Shirly and Dr. Ben Oldfield have developed Spatial Automated Live Sports Audio (SALSA), a method that uses the existing 12 shotgun microphones around the pitch to detect the ball kicks. The system not only looks for overall level intensity, but also the envelope across a range of frequency bands for each sound event type that a sound mixer might want to capture.

Additionally, the SALSA team has developed a number of *acoustic feature templates* that define the range of sound types that sound mixers are looking for and want to capture. For football, these include ball kicks/headers and whistle blows, but there would be different templates for other sports.

When the software detects a match to a template in one microphone, it identifies the same event in other microphones. Then the system does some triangulation based on sound arrival at each mic to derive a line across the pitch along which the ball kick must have occurred. Once these conditions are met, the system can determine with confidence that there was ball kick.

SALSA has a number of acoustic feature templates that define the range of sound types that we are looking for and that we want to capture. For football these include ball kicks/headers and whistle blows. (They'll be different templates for other sports.) So we aren't looking for just overall level intensity but more the content and envelope across a range of frequency bands for each sound event type that we want to capture.

When the software spots a match to a template in one mic it tries to identify the same event in other mics. Then, as you say, we do some triangulation based at sound arrival at each mic to derive a line across the pitch along which the ball kick must have occurred. Once we have one of these for two pairs of mics we can say with confidence the location of the ball kick. SALSA's algorithm is also detecting ball kicks that are virtually inaudible on the mic feeds; it is better at recognizing sound events than our ears are at least.

The SALSA algorithm is capable of detecting ball kicks that are virtually inaudible on the microphone feeds and is more reliable at recognizing sound events than our ears. To make it work accurately and reliably, you keep a high detection threshold to avoid raising faders for inaudible ball kicks.

During live production, SALSA uses one of two approaches. It can automate a mixing console's faders to capture each on-pitch sound event or use the frequency/envelope information of the ball kick to trigger pre-produced samples. These sounds can be added to the on-pitch sounds or can replace the game sounds when the on-pitch capture is poor. If you want it to sound like an EA Sports Game or a Saturday afternoon match on SKY, it is up to you as the sound designer.

This method is already working in the domain of audio objects. Each ball kick and each whistle blow is defined within the software as a short-term audio object with metadata. Each object is tagged with metadata that says what kind of object it is (e.g. ball kick, whistle blow), the sounds duration, and its coordinate location on the pitch. All this is achieved completely independently of camera cuts and the traditional audio follow video which, at the end of the day, really does not work well in most situations.

As a production tool, the SALSA system can provide the sound mixer with separate channel feeds for ball kicks and whistle blows that can be mixed with crowd and commentary feeds, leaving the mix free to concentrate on making it sound awesome instead of chasing the ball with console faders.

SALSA uses one of two approaches: automate a mixing console's faders to capture each on-pitch sound event; or use the frequency/envelope information we have about the ball kick to trigger pre-produced samples that are the closest match to the detected kick. Or if you want it to sound like EA Sports FIFA games, these broadcast sport sounds can be added to the on-pitch sounds or can replace them where on-pitch recordings are poor.

For the first we have to keep a high detection threshold to avoid raising faders for inaudible ball kicks. For the second we can retain a low threshold. In both cases we provide the sound supervisor with separate channel feeds for ball kicks and whistle blows that can be mixed with crowd and commentary feeds, leaving the sound supervisor free to concentrate on making it sound awesome instead of chasing the ball with console faders.

Object-ready audio capture – additionally with this method we are already working in the domain of audio objects, each ball kick and each whistle blow is defined within the software as a short-term audio object with metadata. Each object is tagged with metadata that says what kind of object it is (e.g. ball kick, whistle blow), duration, and its coordinate location on the pitch. At IBC and NAB we collaborated with DTS and Fairlight to show how SALSA could be integrated into object-based audio broadcast. A Fairlight sidecar translated our metadata into DTS's open MDA object audio standard ready for broadcast.[9]

The mixing console is the center of operations for almost all audio functions from localizing objects to spatial processing. There is still a need for large mixing desks with real knobs and faders because of the real-time demands of live entertainment and sports in particular. Virtual audio features as well as touch screens are an integral part of large format consoles, however screens and touch-sensitive controls dominate the studio world with DAW production. Artificial intelligence is quietly creeping into the broadcast world and is a valuable tool for many repetitive tasks, but is not a replacement for qualitative decision making.[9]

Notes

1 alan-admin. 2019. "Pair Wise Amplitude Panning presented by Blumlein in 1931." Alan Blumlein. 2019. www.alanblumlein.com/.
2 Ville Pulkki. 2001. *Spatial Sound Generation and Perception by Amplitude Panning Techniques.* Espoo: Helsinki University of Technology.
3 Knowles, Tom. 2019. Review of Tom Knowles, Solid State Logic Product Manager – Broadcast Systems, Introduces the New SSL System T S500m Interview by SSL Live. www.facebook.com/SSLLive/posts/tom-knowles-solid-state-logic-product-manager-broadcast-systems-introduces-the-n/2696094723753139/.
4 Knowles, Tom. 2019. Review of Tom Knowles, Solid State Logic Product Manager – Broadcast Systems, Introduces the New SSL System T S500m Interview by SSL Live. www.facebook.com/SSLLive/posts/tom-knowles-solid-state-logic-product-manager-broadcast-systems-introduces-the-n/2696094723753139/.
5 Emmott, Kevin. 2016. Review of Calrec Engineering Interviews with Kevin Emmott, Marketing Manager Interview by Calrec Engineering.
6 Lawo, Phillip, Andreas Hilmer, Christian Struck, Jeffrey Stroessner, and Felix Krückels. n.d. Review of Lawo Interviews by Engineering and Marketing.
7 Lawo, Phillip, Andreas Hilmer, Christian Struck, Jeffrey Stroessner, and Felix Krückels. n.d. Review of Lawo Interviews by Engineering and Marketing.
8 Lawo, Phillip, Andreas Hilmer, Christian Struck, Jeffrey Stroessner, and Felix Krückels. n.d. Review of Lawo Interviews by Engineering and Marketing.
9 Interviews with Dr. Ben Shirly and Dr. Ben Oldfield.

4 Immersive Sound Production Using Ambisonics and Advance Audio Practices

Spatialization with Ambisonics Production Methods

It is abundantly clear that across all media platforms 3D immersive sound is critical to the authenticity of high-definition pictures, 360 video, virtual reality and augmented reality. The fact is that multichannel, multi-format audio production is not going away and there should be a commitment by all audio producers and practitioners to advance the quality of audio to the consumer using every tool and practice available.

Ambisonic production is a flexible and powerful creation tool for the spatialization of audio for a convincing immersive experience. Audio producers and practitioners have begun to realize the benefits from the use of ambisonics because of its unique approach to capturing, processing and reproducing the soundfield. Dolby Atmos and MPEG-H 3D support ambisonics, however broadcast and broadband industries have been slow to use and adopt ambisonics as a production platform and tool. Significantly though, in the last few years, ambisonics has been adopted by YouTube and Facebook for 360 video because ambisonics is the only platform that truly and accurately tracks user interactivity with smooth and efficient soundfield rotations from the camera's point of view.

Ambisonic audio production has been around a while, albeit slightly heuristic in the early days, and it was not until significant magnifications of the theory that resulted in Higher Order Ambisonics (HOA) that some of the early ambisonics models became valuable for advanced audio production. Soundfield deconstruction and reconstruction is a powerful tool, particularly because of the flexibility that ambisonics provides with the capability of rendering to a vast range of production and reproduction options. HOA is a far more sophisticated production tool than the proponents of early ambisonics ever envisioned and there are clearly significant advantages to HOA and scene-based delivery over current channel-based and object-based multichannel audio workflows and practices.

What is Ambisonics?

Ambisonics is a completely different approach that as much as possible captures and reproduces the entire 360 immersive soundfield from every direction equally – sound from the front, sides, above and below to a single capture/focal point. Ambisonic attempts to reproduce as much of the soundfield as possible regardless of speaker number or location because ambisonics is a speaker independent representation of the soundfield and transmission channels do not carry speaker setups. Since HOA is based on the entire soundfield in all dimensions, a significant benefit with this audio spatial coding is the ability to create dimensional sound mixes with spatial proximity with horizontal and vertical localization.

DOI: 10.4324/9781003052876-4

How Does It Work?

A propagating sound wave originating from one source does not move in a straight line but expands in a series of sphere-shaped waves equally in all directions. The behavior of sound waves as they propagate through a medium and even how sound waves reflect off an object was explained by the principle of wave fronts by the Dutch scientist Christiaan Huygens. A wave front is a series of locations on a sound wave where all points are in the same position on that sound wave. For example, all points on the crest of the same wave form a wave front. Huygens further states that each point on an advancing wave may be considered to be a new point source generating additional outward spreading spherical wavelets that form a new coherent wave.[1]

Expanding functions of a sphere and soundwave expansion can be explained by spherical wave fronts that may vary in amplitude and phase as a function of spherical angle and can be efficiently modeled using spherical harmonics which are mathematical functions (models) for mathematical analysis in geometry and physical sciences. Spherical arrays can be used for soundfield analysis by decomposing the soundfield around a point in space using spherical harmonics.

Decomposing a soundfield to spherical harmonics is a process of converting the soundfield to Associated Legendre Polynomials which map the angular response and Spherical Bessel

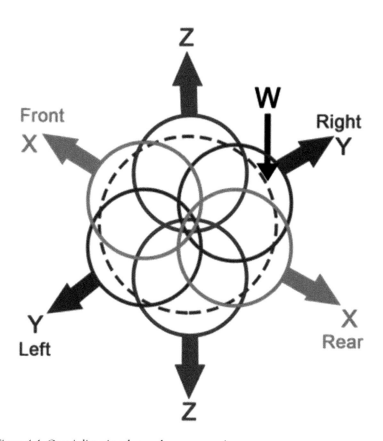

Figure 4.1 Omni directional soundwave expansion

Functions which map the radial component and together form the Spherical Harmonic functions. HOA coefficients are the coefficients that are used to formulate the desired combination of the spherical harmonics functions. Spherical harmonic functions are important to spherical coordinates and solving equations with wave propagation and integral to calculating HOA coefficients.

Soundwaves can create a complex soundfield that can be composed of hundreds of sound sources creating their own sound waves, diffractions, scatterings and reflections. Ambisonics is a method to capture as much of the 3D soundfield as desired up to the maximum HOA order, and captures both direct and reflected sounds. Considering that sound comes at us from every direction it was not a giant leap to consider capturing the entire soundfield at a point in space from a single point receptor – a microphone or listener.

Reproduction of soundfields begins the synthesis of a soundfield by recombining the amplitudes of the spherical harmonics and making the reproduced sound match the measured soundfield.

Ambisonics was presented by Dr. Michael Gerzon based on psychoacoustical consideration where he developed a mathematical model for capturing and reproducing a simple dimensional soundfield. First generation ambisonics was and is a 3D format although a low-resolution format that never caught on outside of the UK till VR's adoption of the format.[2]

The spatial resolution for basic ambisonics is quite low but can be increased by adding more direction components to achieve a more useable ambisonic format called HOA (Higher Order Ambisonics). HOAs are based on a mathematical framework for modeling 3D soundfields on a spherical surface where the HOA signals can be calculated based on the spatial location of the sound sources. The HOA signals can be derived from spatial sampling and spatial rendering the three-dimensional space. HOA is used to reconstruct the dimensional soundfield by decomposing the soundfield into spherical harmonics which contain spatial information of the original soundfield. Significantly HOA signals preserve the spatial audio information.

The soundfield modeling projects the soundfield onto a set of spherical harmonics, and the number and shape of the spherical harmonics determine the resolution of the soundfield. HOA project/inject more spherical harmonics into the equation. Spherical harmonics are special functions defined on the surface of a sphere. Each additional harmonic coefficient adds higher spatial resolution to the modeled or synthesized soundfield.[3]

Spherical Harmonics - 3rd Order

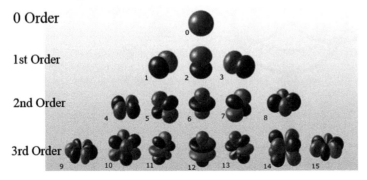

Figure 4.2 Spherical harmonics – 3rd Order

What is Scene-Based Audio?

Ambisonic audio production is nothing new but clearly is a paradigm shift in the contemplation and production of future ready sound. One reason ambisonic lagged in acceptance was the fact that ambisonics' true benefits were not realized till the expansion of the original concept to higher-order-ambisonics along with the development of powerful production tools. Typically audio production tools have been compressors, dynamic controllers and equalization, but spatialization of audio has continued to develop past reverberation and room simulation.

The capability to construct high resolution soundfields depends on the ability to capture, construct and render soundfields with the highest possible resolution. Proponents of HOA have titled advance ambisonic production SBA – scene-based audio. Scene-based production is the natural evolution of HOA although now with the tools to increase the flexibility of HOA, but at the end of the day it is still a HOA production with advanced scene-based tools. HOA is a process and SBA are the tools to make HOA more useful as a production tool.

Scene-based audio has an enhanced set of audio features for manipulation of the audio scene during playback. It provides the user the flexibility to alter POV – points of view, zoom or focus on a specific direction, mirror, attenuate or amplify an audio element as well as rotate the soundfield. Unique about ambisonic production is the ability to deliver any combination of visual experiences – TV, VR and 360 video on a single audio workflow.[4]

How Does Ambisonics Operate?

The concept is simple – capture the entire soundfield then render and reproduce as much of the soundfield as possible. Ambisonics looks at the soundfield as a grid of equally spaced sound zones that need to be captured to a single point.

This macro-level approach can be derived from a combination of mono stereo and multi-capsule array microphones similarly to the way broadcasters capture sound today. The problem with all microphone capture is that the further the microphone is from the source the more diluted the signal. The inverse square law states that the intensity of the sound will decrease in an inverse relationship as the soundwave propagates further from the sound source. For every factor of two in distance, the intensity of the soundwave is decreased by a factor of four. Additionally, microphones not only capture the sound you want but also a lot of what you

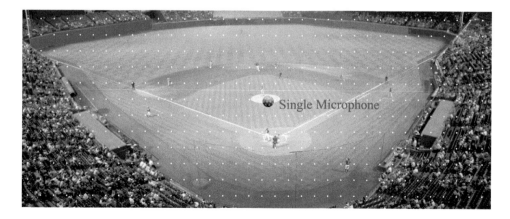

Figure 4.3 Higher order ambisonics (HOA) listens to the entire soundfield at a macro level

do not want – background noise. It is difficult to isolate objects and estimate their individual metadata.

Microphone position is never optimal in sports. Building on the concept of entire soundfield capture and that ambisonics treats sound from all direction equally, then microphone placement at baseball would have to be over the pitcher to capture the infield equally and deliver an immersive crowd. Since this is an impossible location to achieve, multiple microphone positions are located to capture a balanced, holistic representation of the complete soundfield. Additionally, closely correlated microphone arrays lack much capture detail beyond a relatively small sound capture zone ultimately requiring additional "spot" microphone for detail.

A significant practical aspect of HOA is that you do not need to fully capture in HOA, but can optimize individual microphones into HOA. Individual microphones can be placed symmetrically or separated at arbitrary locations and use the capture information from all of these microphones to derive the HOA coefficients. The HOA encoder generates the 3D coefficients. However, when you capture an ambisonics foundation it will deliver desirable and predictable results for you to build your detailed and specific sound element on top of this foundation.

Multiple HOA capture points have been suggested and as the costs come down and the performance of multi-capsule ambisonic microphone improve, multiple ambisonic microphones may be realistic. Significant is the fact that previously produced and up-produced music and legacy content that is processed into HOA material can be added to an HOA mix.

Encoding HOA creates a set of signals that are dependent on the direction and position of the sound source and not the speaker for reproduction. With the audio rendered at the playback device, the playback renderer matches the HOA soundfield to the number of speakers (or headphones) and their location in such a way that the soundfield created in playback resembles closely that of the original sound pressure field.

A typical scene could contain hundreds of objects which with their metadata must be recreated on the consumer device. Not all consumer devices are created equal and may not have the ability to render complex scenes. Significantly HOA's rendering is independent of scene complexity because spatial characteristics are already mixed into the scene.

Figure 4.4 Higher order ambisonics (HOA) multiple capture zones

Production Note

HOA allows the sound designer/producer to create or develop the direction of the sound and not be tied to where the speaker or reproduction device may be, which is contradictory to the way a lot of sound is produced.

The fact is that dimensional sound formats will not go away. People want options for consumption and the challenge is how to produce and manage a wide range of audio experiences and formats as fast and cheaply as possible. Cheaply also means minimizing the amount of data being transferred. If the rendering is done in the consumer device, it inherently means that there is a need to deliver more channels/data to the device. However, data compression has significantly advanced to the point that twice as much audio can be delivered over the same bit stream as previous codecs.

Now consider the upside of the production options using HOA. You have the ability to reproduce essentially all spatial formats over 7 of the 16 channels (a metadata channel is needed), then you have another eight individual channels for multiple languages, custom audio channels, and other audio elements or objects that are unique to a particular mix.

Additionally, producing the foundation soundfield separately from the voice and personalized elements facilitates maximum dynamic range along with loudness compliance while delivering consistent sound over the greatest number of playout options.

The sonic advantages of ambisonics reside with the capture and/or creation of HOA. Ambisonics works on a principle of sampling and reproducing the entire soundfield. Intuitively, as you increase the ambisonic order the results will be higher spatial resolution and greater detail in the capture and reproduction of the soundfield. However, nothing comes without a cost. Greater resolution requires more soundfield coefficients to map more of the soundfield with greater detail. Some quick and easy math: fourth order ambisonics requires 25 coefficients, fifth order requires 36, and sixth order requires 49 and so on. The problem has been that HOA Production requires a very high channel count to be effective which did not fit in the current echo system, but coding from Qualcomm and THX has reduced the bandwidth for a HOA signal to fit in 8 channels of the 15 or 16 channel architecture leaving channels for objects and interactive channels.

Dr. Deep Sen has been researching the benefits of HOA for decades and headed a team that developed a "mezzanine coding" that reduces the channels up to the 29th order HOA (900 Channels) to 6 channels + control track. Now consider a sound designer's production options. HOA provides the foundation for stereo, 5.1, 7.1, 7.1+4, 10.2, 11.1 up to 22.2 and higher using only 7 channels in the data stream. I suspect that there are points of diminishing returns. Scalability – the first four channels in a fifth order and a first order are exactly the same.[4]

Capture or Create?

HOA is a post-production as well as a live format; however, live production is dependent on complexity and latency.

High Order Ambisonics: HOA in Post-Production

3D Panning Tools, effects and ambisonics mastering tools reside in the editing and hosting platform of Ableton while Neuindo and ProTool use third-party plug-ins for immersive sound production.

Ableton includes 3D positioning tools, azimuth and elevation, 3D effects and masters in real time to HOA. Each channel can have a set of tools. Ableton includes a couple of interesting programs including a spinning program that automates motion in 3D space with vertical rotations, convolution reverbs and three-dimensional delays from a single channel. Ableton's output is saleable from headphones to immersive speaker arrays.

Neuindo is a popular DAW that supports dearVR immersive, allowing the sound designer to create immersive and 3D content. For an action sound designer the Doppler Effect plug-in does a nice job simulating the perception of movement and distance by pitch changes as the source passes you.

ProTools HD is a widely used DAW in post-production, however it derives much of its functionality from third-party plug-ins. A set of scene-based HOA Tools was developed under the guidance of Dr. Deep Sen and resulted in significant advancements for further production and development for HOA.

Because HOA deals with spherical expansion, tools like rotation, sphere size and interesting room and reverb simulation programs have been developed. Distance is interesting because you are not just changing the volume when you move a sound element closer or farther away, but as in the real world the change in distance can change the tone of a sound as well.

The ability to adjust the size of an object has fascinating production possibilities. Size expands the perception of magnitude of a sound element by diverging the sound element into adjacent channels.

Size is a processing feature that can be useful in speech intelligibility or as an effect for dramatic enhancement. The soundfield can also be widened or squeezed to match the TV size. Building acoustic environments is common with object-based audio production but spatial enhancements have proven effective in immersive sound production for both speaker and ambisonic methods of production as well. Room simulators are capable of creating acoustic space and use complex reflection algorithms to recreate the variety of a dimensional space. The ability to contour parameters such as reflections and diffusion empowers the sound designer in recreating and creating the realistic sonic space.

Facebook offers a free Audio 360 Spatialiser plugin that replaces the conventional DAW panner giving each channel full 3D positioning, distance and room modeling. The channel input options are mono, stereo, 4.0, 5.0, 6.0, 7.0 and B Format 1st, 2nd and 3rd Order ambisonics, as well as controls for azimuth, elevation, distance, spread, attenuation, doppler, room modeling and directionality. Ambisonic controls are source roll, pitch and yaw, plus the ability to control the diffuse soundfield.

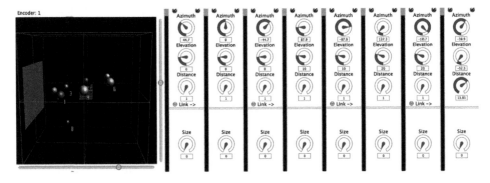

Figure 4.5 Channel control strip from a HOA tools plug-in with controls for azimuth, elevation, distance and size

The focus effect is not baked into the mix and can be controlled from your app in real time or encoded into a FB360 video as metadata. Focus control gives the sound designer the ability for a range of controls from full headtracking to the ability to define a mix area and have sounds outside that zone attenuate. Focus control includes focus size of the area and off-focus level, the attenuated level outside of the focus area. Focus azimuth/elevation are values that are relative to the listener's point of view when the headtracking is disabled. In the loudness plugin the mixer can set the overall loudness of your mix and the maximum loudness and true peak in your mix as if the listener were looking at the loudest direction.

Facebook describes Audio 360 as an immersive sphere of audio and is tied to headphones. 360 video can be viewed on screens and goggles. Facebook 360 provides a suite of software tools for publishing spatial audio and video that can be exported to a range of formats including ambisonics. The format sports 4K resolution, image stabilization, VR and can stream live, making possible unique entertainment experiences.

A feature I found unique is called "Points of Interest" which is a unique production tool to guide your viewer through your video. Ambisonics is the only format that locks the picture to the sound for rotation and more.

Many of the Ambisonic Tools are 1st Order and are often mastering tools like the Waves B360 Ambisonic Encoder which has panner-like controls then outputs the channels to four channels of B Format with gain and phase information equivalent to its direction in the soundfield. Additionally, YouTube video supports 1st Order ambisonics with head-locked stereo.

High Order Ambisonics (HOA) – Live

The first broadcast of MPEG-H using HOA was done by 19 different manufacturers at the 2018 European Athletics Championship in Berlin. The test demonstrates an end-to-end production workflow – capture, process, record and distribute live UHD content featuring high frame rates and dynamic range with Next Generation Audio. The tests used familiar workflow using a combination of mono, stereo and multi-capsule array microphones similarly to the way broadcasters capture sound today. The SSL mixing console supports height channels and the panning of the microphones was accomplished on the mixer. The mix output was encoded and streamed.

Complex HOA production will probably need processing, which results in latency. As of publication, a significant problem for live capture using multichannel array microphones above 3rd Order is that the microphones are computer-controlled arrays with advance processing that may have too much latency for exact lip-sync. It appears that 1st and 2nd Order ambisonics can be used with whatever amount of latency comes with the format conversion or decoding, but 3rd Order and greater ambisonics appear to require more processing, resulting in more latency and more problems.

A sporting event where the crowd was reproduced with a few frames of delay would probably be perfectly acceptable. Up-producing music using HOA would certainly result in some latency during the up-conversion, however would have no detriment to the production. A soundfield foundation with a static capture and reproduction will result in minimum latency.

HOA technology fulfills the need to produce immersive content, distribute it in an efficient bit stream and have it played out on a wide variety of speaker configurations and formats – creatively, efficiently and in a consumer-friendly way. By simply rendering the underlying soundfield representation, HOA ensures a consistent and accurate playback of the soundfield across virtually all speaker configurations. This process seems to work well across many playback formats and could possibly eliminate the need to downmix or upmix to achieve anything from stereo to 22.2 or more. This concept could be a significant solution to a problem that has burdened sound mixers who have to produce both in stereo and surround.

Spatialization: Advance Audio Production Techniques and Practices (AAPTP)

Advanced audio production techniques are beyond the room reverbs and echo-type devices of earlier times. Mixing consoles provide the basics for dynamics management, tone control and fundamental panning, but advance spatial enhancement is done with applications and processes using plug-ins and out of mixing console processors both in live and post-production workflows. Plug-ins are a specific application that can be added in the production signal chain and are usually hosted and resident in the mixing console. Before plug-ins there has been a history of using standalone "blackboxes" for signal processing that were patched into the signal flow to process the audio.

With the migration from analog mixing consoles to digital mixing desk came the possibility of advance signal processing inside the digital desk. All digital mixing consoles contain equalization, time shifting and dynamics processing designed and built in by the manufacturer, but until recently there has been a reluctance for console manufacturers to unlock the proprietary audio engine to third-party application developers. For the manufacturers there was a higher comfort level with a side chain, blackbox-type device as opposed to an in-line application crashing and shutting down the mixer.

All audio console manufacturers discussed in this book have integrated third-party applications into their mixing platforms and this will continue to advance. However, you should always exercise caution when adding any new application to a computer platform and remember all computers crash at some time. Additionally, always listen for latency and digital artifacts that will affect the clarity and quality of your sound.

Advance audio production should be looked at as an umbrella of tools that not only can adjust the spatial properties of sound elements, but also change the tone and sonic characteristics of spaces. Basic spatialization can be as simple as time and timbre difference between a direct sound and a delayed or diffused element of the original sound. This is what is known as basic reverb or echo and occurs naturally from reflections off surfaces in the path of the original sound waves. This basic tool can simulate a concert hall or the natural spatialization of sound like what you hear when you are in an expansive European Cathedral. Our brain tells us this is a cathedral and there is an expectation of what the sound of a cathedral is.

Virtual simulators are a growing theme of plug-ins that can create and shape any sonic characteristics of a sound element including size, magnitude and distance, as well as adjust the spatial characteristics of a sonic enclosures where a sound object resides. Advance Audio Production uses advance modeling and virtual simulation done on plug-ins and hosting computers and depending latency can be done in real time and applied live. A composite soundfield is often an amalgam of sound layers that have been spatialized to complete the dimensional soundfield which can be forgiving with precise synchronization and localization.

Where Did All This Come From?

Hearing and the perception of sound is uniquely personal. Many factors affect our hearing including the shape of our head, ears and the physical condition of our auditory system. These factors impact the natural collection of sound by humans just as the electrical, mechanical and physical characteristics of microphones effect the quality of sound collection.

Beyond the physical collection of sound is the processing and interpreting of sonic information. Psychoacoustics is the science of how the human brain perceives, understands and reacts to the sounds that we hear. Perception of sound is affected by the human anatomy, while cognition is what is going on in the brain.

Limits of perception

The human auditory system can only process sound waves within a certain frequency range. This does not mean these extended frequencies do not exist, just that humans do not process them through the auditory system. Additionally, the auditory system does not process all frequencies the same. Some frequencies are more intense even when they are at the same amplitude. For example, low frequency sound waves require significantly more energy to be heard than high frequencies. Our hearing is not linear and the equal loudness curves known professionally as the Fletcher-Munson curves show the relationship between frequency, amplitude and loudness. Finally, complex soundfields can suffer from frequency masking. Two sounds of the same amplitude and overlapping frequencies are difficult to understand because the brain needs a minimum difference in frequency to process the sound individually.

Sound localization is impacted by the size of the head and chest and the physical distance between the ears. This is known as the head related transfer function (HRTF). The sound usually reaches the left ear and right ear at slightly different times and intensities and along with tone and timbre the brain uses these clues to identify the location a sound is coming from.

Cognition is what happens in the mind where the brain infuses personal biases and experiences. For example, when a baby laughs there is one reaction as opposed to when they cry. Cognitive perception of sound has created an entire language of sound. Defining and describing sound is often a difficult exercise because our vocabulary for sound includes descriptive phrases that comprise both objective and subjective metaphors. Technical characteristics such as distortion, dynamic range, frequency content and volume are measurable and have a fairly universal understanding, but when describing the aesthetic aspects and sonic characteristics of sound, our descriptors tend to become littered with subjective phraseology. Here is the simple yet complicated phrase which has always irritated me: "I don't like the way that sounds." Exactly what does that mean? I worked with a producer who made that comment halfway through a broadcast season during which I had not changed anything substantial in the sound mix. Being the diligent audio practitioner, I took his comment to heart and really spent time listening to try to understand why he said what he said.

Broadcast sound and sound design is a subjective interpretation of what is being presented visually. The balance of the commentary with the event and venue sound is interpreted by the sound mixer. The sports director and producer are consumed with camera cuts, graphics and replays while possibly focusing on the sonic qualities of a mix may be beyond their concentration. Factor in the distractions and high ambient noise levels in an OB van – remember technical communications are verbal and everybody in the OB Van wears headsets – and now you have to wonder who is really listening.

Meanwhile, after objectively listening and considering what the problem could be, I inquired about the balance of mix, its tonal qualities, and my physical execution of the mix. Once again the answer was, "I don't like the sound." My next move was to look really busy and concerned and ultimately do nothing. That seemed to work.

When surround sound came along, a common description emerged to describe the sound design goals: to enhance the viewer experience. At least now when there is talk about multichannel 3D sound, the conversation begins with the nebulous notion of immersive experience. I think this has to do with creating the illusion of reality … go ahead, close your eyes … do you believe you are there?

So what do balance, bite, clarity, detail, fidelity, immersive experience, punch, presence, rounded, reach, squashed or warmth have to do with sound? As audio practitioners we seem to act like we know. After all, we make that mysterious twist of the knob and push of the fader achieve audio nirvana, but audio descriptors are important to humanize the audio experience and conquer the psychoacoustic and physiological aspects of sound.

The psychology of sound also has to do with the memory of sound and reminders from physical cues such as pitch, frequency, tempo and rhythms triggers a sensory and perhaps emotional experience. I believe that if you have ever heard a beautiful voice or guitar then that becomes the benchmark for reference. A lot of what a sound designer has to do is satisfy the memory, but I argue perhaps it is time to create a new impression. Psychoacoustics could be considered how the mind is tricked by sound while the physiological aspects of sound reinforce the illusion. For example, low frequencies, a fast tempo or pace affect breathing and cardiovascular patterns. When I mixed car racing I always tried to emphasize the low frequencies of the cars to heighten the visceral experience.

Principles of Psychoacoustics

Understanding how we hear, along with how the brain perceives sounds, gives sound designers and software engineers the ability to model sound-shaping algorithms based on psychoacoustic principles and thought. Common considerations when modeling sound are frequency and time, so instead of using elevation to achieve height try using equalization which can be an effective means for creating impression of height. We naturally hear high frequencies as coming from above because high frequencies are more directional and reach our ears with less reflection. This principle is known as the Blauart Effect.[5]

Significantly, a lot of the low frequency energy has already been lost. By equalizing certain frequencies, you can create the illusion of top and bottom; in other words, the greater the contrast between the tone of the top and the bottom, the wider the image appears to be. This principle works well for sports and entertainment because you can build a discernable layer of high frequency sounds (such as atmosphere) slightly above the horizontal perspective of the ear.

The Haas Effect/The Precedence Effect

The Haas Effect/the precedence effect is the founding principle of how we localize sounds. The Haas Effect, also known as the precedence effect, is a key psychoacoustic principle that can be applied to create the illusion of width and a realistic sense of depth and spaciousness. Helmut Haas explained why, when two identical sounds occur within 30 milliseconds of each other, the brain perceives the sounds as a single event. Depending on frequency content this delay can reach as much as 40ms. Short delays result in the signal going in and out of phase and are underlying concepts for chorus, flanger and phase types of devices that are not used in broadcast but proper application creating a wider perception of space is beneficial to the sound mix.

Blauert came to the same conclusion as Haas about delay and localization, in that as a constant delay is applied to one speaker the phantom image is perceived to move toward the non-delayed signal. Blauert further said that the maximum effect is achieved when the delay is approximately 1.0ms.

Because the ears can easily distinguish between the first impression of a sound and its successive reflections this gives us the ability to localize sound coming from any direction. The listener perceives that the direction of the sound is from the direction heard first – preceding the second. While panning manipulates the sound by affecting the levels between the left and right channel, the Haas effect works because of the timing difference between the channels exactly the way our ears work. The precedence effect helps us understand how binaural audio works as well as how reverberation and early reflection affect our perception of sound. [6]

There have been some studies about how our perception of sound changes with a change in sound characteristics such as pitch shift or frequency variation. The Doppler shift is a valuable audio tool to enhance the sense of motion. It has an additional effect that appears to move or shift high frequencies above the listener. The faster a sound source is, the higher the sound is pitched up. The Doppler shift can be captured live with microphone placement, however there are some programs that can effectively emulate this effect.[7]

Phantom Imaging – Virtual Sources – Phantom Sources

All channel-based reproduction systems, such as stereo, surround and immersive, produce phantom imaging where we perceive a sound source between channels/speakers from level and time interactions.

Psychoacoustic Masking

Psychoacoustic masking is the brain's ability to accept and subdue, to basically filter certain distracting sounds. I have read articles touting the miracles of sound replication by Edison's Victrola. Edison demoed real singers side by side with his devices, he would pull back the curtain exclaiming better than life, pay no attention to those pops and ticks in the recording. The mechanical reproduction devices suffered from a significant amount of scratches and ticks but the brain filters out the undesirable noise. For example, radio static is filtered out by the brain when a high proportion of high frequency components are introduced. Additionally, noise and artifacts from over compressed digital sampling may be filtered by the brain but result in unhealthy sound.

The Missing Fundamental Frequency

The missing fundamental frequency is an acoustical illusion resulting in the perception of nonexistent sounds. The harmonic structure determines our perception of pitch rather than strictly the original frequency. The brain calculates the difference from one harmonic to the next to decide the real pitch of a tone even when the fundamental frequency is missing. This is the reason why you can hear sounds over small speakers that cannot reproduce the full range of frequencies – the brain fills in the missing fundamental frequency. Sub-harmonic synthesizers create the tone as a virtual pitch below the audible frequencies of hearing.

At certain frequencies harmonics in the mix can contribute to the boosting of certain frequencies. Additive spectral synthesis can be used for adjusting the timbre of your sounds by combining and subtracting harmonics.[8]

Applied Psychoacoustic: Creating Production Tools

The physics of an environment, the ear and the brain are at play when creating psychoacoustic production tools. Acoustic simulators in the 50s were as basic as spring reverbs and it was a time when stereo widening was achieved by adjusting the relationship of the sides and the center signal. But no more. Some manufacturers and researchers go into a variety of halls and spaces and do impulse measurements of decay times, reverberation field measurements and vector analysis of reflections to try and mimic the real soundfields.

3D audio effects involve the virtual placement of sound anywhere in front, to the sides, behind and above the listener. Spatial enhancements such as reverb and room simulators are useful tools in dimensional and immersive sound production because they recreate the perception of the physical size of a space as well as playing a significant role in creating the illusion of a three-dimensional space.

Basic reverb and delays are a single dimension balance between the direct sound and reflected energy where advance audio production techniques are three-dimensional, founded on psychoacoustic considerations. Spatialization can also be achieved by processing an audio signal and by infusing the processed signals into the immersive soundfield. There are room simulators as well as a variety of dimensional reverberation programs that can effectively process an audio signal into a variety of immersive formats with height control. This type of processing gives cohesion between the lower and upper layers as well as control of the reflections and diffusion of the returning audio signals.

Psychoacoustic modeling software can take a sound or group of sounds and digitally recreate them in a digital acoustic map of essentially any desired sonic space – virtualization of space. Room simulators are capable of creating acoustic space using complex reflection algorithms to recreate the variety of a dimensional space. The ability to contour parameters like reflections and diffusion empowers the sound designer in recreating and creating the realistic sonic space.

Advanced Spatial Enhancement

In addition to panning and placement, spatialization tools are capable of distance and size functions. Distance is interesting because you are not just changing the volume when you move a sound element closer or farther away, but as in the real world the change in distance can change the tone of a sound as well. Size is a 360-degree hemispherical assessment or perceptual evaluation of expanse, and advance spatial enhancement tools can expand the apparent dimensional aspects or size of a sound element beyond the original region enhances the dimension or magnitude of the original sound element.

The ability to adjust the size of an object has fascinating production possibilities. Size expands the perception of magnitude of a sound element by diverging the sound element into adjacent channels. Size is a processing feature that can be useful in speech intelligibility or as an effect for dramatic enhancement. The soundfield can also be widened or squeezed to match the TV size. In short, there are many tools that are added to the creative tools of the audio mixing engineer.

The Secret Sauce: Plug-Ins and Other Black Boxes

Digital mixing desks and digital audio workstations (DAWs) depend on plugins for increased functionality and expansion. Beyond localization environment/room simulators is a valuable tool in the advance audio toolbox.

Creating an immersive soundfield for outdoor winter sports is challenging because, in reality, wind does not make any sound until it collides with something like the trees in a forest. I created several of these challenging soundscapes underscoring that the ability to create such believable soundfields is a powerful live production tool. The DSpatial audio engine can operate in a standalone configuration using, or being used to generate, a soundfield in real time.

DSpatial

DSpatial created a bundle of plug-ins that work under the AAX platform in a fully coordinated way. Reality builder is inserted on each input channel and can operate in real time in the pro tools environment along with the option to run off-line. In off-line mode the rendering is much faster than in real time. The DSpatial core engine can run in stand-alone mode which means no latency.

I enjoyed a discussion about sound design principles and practices with Rafael Duyos, the brains behind DSpatial, who is much more than a coder. I believe he gets what sound designers dream of.

DENNIS BAXTER (dB): As a sound designer creating a sense of motion and speed has always been a challenge, particularly with events that do not have a lot of dynamics like downhill skiing or ski jumping. Creating the illusion of someone flying through the air on a pair of skis is a challenge.

RAFAEL DUYOS (RD): Scientifically speaking, what we have done is a balanced trade-off between physical and psychoacoustic modeling principles. By that I mean that if something mathematically correct doesn't sound right, we have twisted it until it sounds right. After all, film and TV are not reality but a particular interpretation of reality. So we are not always true to reality, but we are true to the human perception of it.

RD: We have applied this (principle) to all the effects we have modeled. For example, Doppler is a direct consequence of the delay between the source and the listener, when any or both are moving in relation to the other, but we have made this delay optional because sometimes it can become disturbing. Inertia was implemented to make the Doppler effect more realistic by simulating the mass of moving objects. Inertia is applied to each source according to its actual mass. Small masses have much more erratic movements. The Doppler of a fly doesn't sound the same as the Doppler of a plane. Doppler and Inertia usually have to be adjusted in parallel; very high degrees of Doppler usually require more inertia.

In the case of proximity, for example, we have even provided for an adjustment of the amount of proximity effect, from nothing (like current panning systems) to fully realistic. We use equalization only marginally. Normally we use impulse responses and convolutions because they are much more realistic. A very important part of the algorithm is the reflections. Take binaural, for example. A loose HRTF usually doesn't sound very realistic. However, if you take a good binaural microphone, it sounds much better than an HRTF alone, and that's because with the microphones you get millions of micro reflections coming from everywhere. That's what we try to model as much as possible. We are probably the system that needs the most computation to work, but we are not worried about that because computers are getting more and more powerful. Time is on our side.

dB: I thought your program for Walls and doors – reflection, refraction, diffraction and scattering produced by walls and doors was very clever and useful. Can you explain your scatter principle?

RD: Dispersion is achieved through extreme complexity. The key to our system is our Impulse Response creator. This is something that cannot be achieved with algorithmic reverberations, and allows us to get the best of convolution and the best of algorithms.

RD: The complexity of IR modeling allows us to create fully decorrelated IRs for each of the speakers. That's simply not possible with microphone recorded IRs. For us it's the essential part of our design. Our walls, doors, reflection, refraction, diffraction and scattering base their performance on the complexity. Rotate, collapse, explode, etc. are created in our DSpatial native format, and can then be exported to any format, be it Ambisonics, binaural, ATMOS, Auro3D. There is no format limit. As we record the automations and not the audio, we can always change it later.

dB: What are the X, Y and Z controls for?

RD: There is an X Y Z slider for each of the sources, and these represent the positions of that source in the 3D space. As simple as that. If the final mix is not in 3D, the projection of the 3D space in two or one dimensions are accomplished. It is possible to edit on a two or three-dimensional plane or even on an Equirectangular plane. You will automatically see the effect of these movements on the X, Y and Z sliders.

dB: Some of the controls are for Center Force and Space Amount – please explain.

RD: Center Force is a feature that a Skywalker engineer asked for when we showed them our first prototype. They were obsessed with the dialogue being attracted to the center speaker. Somehow Center-Force defines the intensity of attraction that the Centre speaker exerts over the dialogue, as if C was a magnet.

dB: Can you explain Ambients? Is this like an Ambients noise generators?

RD: It is that and much more. Ambients are an audio injection system based on a player of audio sound files, for diverse use. Its first use is to create sounds of environments such as noise from cities, sound from restaurants, parks, people, animal, machines or any general sound environment. Even synthetic sounds made with synthesizers, musical instruments and phrases. In a word: any sound that can be put in an audio file.

RD: Once the type of Ambient is set, it can be injected into the final mix using three-dimensional spatialization parameters, through a simple joystick-like pad that is fully automatable. In addition to ambient sounds, you can use music and sound effects such as

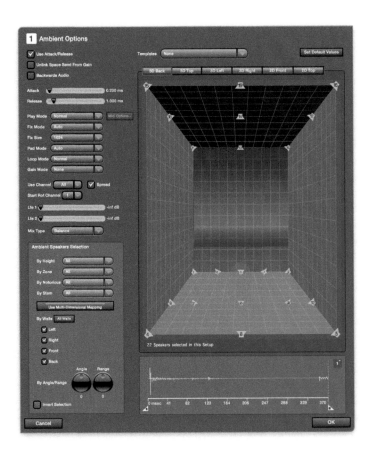

Figure 4.6 DSpatial ambient options window

Figure 4.7 DSpatial-Reality-2-0

gunshots, screaming, horses, door closing sounds, footsteps, etc. In these cases, there is a pad-controlled firing mode, of course, supporting spatialization parameters.

The Ambient system is also intelligent enough to use multichannel audio in both the ambience source and the number of channels in the final mix, ensuring the best possible spatialization.

dB: Can you explain Spatial objects?

RD: Spatial-Object is what we call DSpatial objects which is the next generation of objects. Traditional objects are simple mono or stereo files located in a grid of speakers. They lack the ambience, which in reality is closely linked to the original signal. The environment is supplied separately in the form of beds. But that has the problem that the beds don't have good spatial resolution. If our goal is to make the system realistic, using beds is not a good idea. To be realistic, objects have to be linked to their reflections. But for that you need an integrated system that manages everything. That is exactly what Reality Builder does.

RD: DSpatial-Objects are devoted to production, not just delivery. Contrary to all object-based formats, DSpatial work with objects from the very beginning of a production.

dB: Remember, Dolby required a bed to get started.

RD: With a DSpatial workflow it is ideal is to work dry, and add as much, or as few, reverbs as you want afterwards. There is no need to record the original reflections, hyper-realism and repositioning possibilities DSpatial extreme realism allows for total control in post-production.

This author listens and mixes in a neutral acoustic environment using ProTools, Nuendo and Reaper with 11.1 Genelec speakers 7.1.4 and has auditioned and mixed the plug-ins described in this book.

The ability to create sonic spaces in real time is a powerful tool in immersive sound creation and production. Remember sports sound design is equal parts sports specific, event specific and venue specific. As discussed in Chapter 5, microphones capturing sports specific sound is possible, but capturing the right venue tone is complicated by poor acoustics and little noise control. Advance audio production practices advocate manufacturing an immersive soundbed to develop upon.

Advance audio production practices can be extended to include the aural re-assembly of a hostile acoustic environment where the background noise completely overwhelms the foreground. Such was the distraction with the vuvuzelas at the 2010 World Cup. As I have said, a sports venue has a rather homogenous sound throughout and infusing a new room tone on the venue, similarly to what is done in film, solves a lot of problems.

Sound Particles

You probably have heard sound particles on film-type productions, but sound particles has developed an immersive audio generator that produces sounds in virtual sound worlds. Sound particles is a 3D native audio system that uses computer graphic imagery (CGI) (modeling) techniques to create three-dimensional images in films and television. Sound particles uses similar CGI computer modeling principles to generate thousands of 3D sound particles creating complex sound effects and spatial imaging. All sound particles processes require rendering.

Practice application – sound particles is a post-production plug-in but because of flexible I/O configurations a timed event could be triggered, exported from the live domain to sound particles, rendered and played out live through the sound I/O with the live action. For example, a wide shot of the back stretch of a horse race is probably a sample playback and the sample playback could be processed, rendered in real time and timed to the duration of the horses run along a particular distance.

Sound particles can be anything from a single simple particle to a group of particles forming complex systems. To build a new project from scratch, open the menu and select EMPTY which opens a blank timeline. Now you can build your new timeline with video at the top and then add audio track(s), add particle group, add particle emitter, add microphone or begin with presets.

An audio track is the sound that is going to be processed and can be mono, stereo or ambisonic. This is usually some file format such as a .wave or other audio file. You import your audio file or files to the timeline. In the case of using multiple files each particle will randomly select an audio file from the selection of imported files.

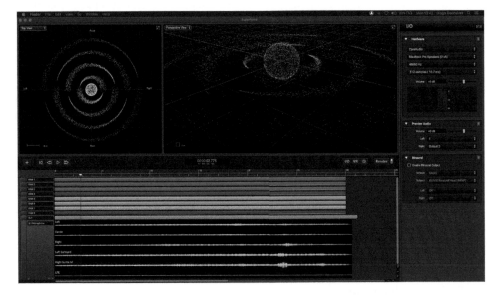

Figure 4.8 Sound particles menu SuperNova

In the menus you can select a particle group where particles start at the same time or a particle emitter that emits particles at a certain rate. In a particle group you can set the number of particles – the default is 100 but the user can select from 1 to 100,000. You can change the shape of the particle, for example circle, cylinder, rectangle or sphere.

Menus provide information about when and where the particle starts after its initial value. Point is when all particles are created at the same point. Line is when all particles are created within a line segment, inside circle, inside rectangle, inside sheer, outside sphere, inside cylinder and cylinder surface.

Movement modifiers control straight line and rotational acceleration.. For example, straight line movement is where each particle is moving in a straight line with gradually increasing or random velocities. While rotational acceleration controls the movement of a particle around its axis, additional menus control audio modifiers such as gain, EQ, time and pitch and delay. An interesting feature is a random delay where each particle will start with a random delay of up to five seconds.

Hundreds of presets for positional automation such as Doppler, explosion, flyby, hurricane, jumping around, machine gun, magnetic poles, moving tunnel, rotating grid, spinning and more can be selected and added to the timeline or automation can be programed by the user.

In order to render the scene you need to have a point of reference – the program uses the concept of a microphone and can be any polar pattern from mono, stereo, multichannel – immersive Dolby Atmos, Auro 3D, NHK 22.2 or ambisonics up to 6th Order. The microphone renders each particle in terms of their distance by attenuating the sound, in terms of direction by applying panning and Doppler effect. You can change the position of the microphone as well as the group on the axis grid.

There are menus for speaker setup from immersive, surround to an edit mode using azimuth and elevation as well as Audio Hardware I/O. Binaural monitoring can happen on any type of audio format and with ambisonics you can have head tracking if you add an ambisonic microphone to the scene.

Render can be online or offline depending on the complexity of the scene. You can render a project with more than one track and more than one microphone. Export the file with interleaved and non-interleaved where the channel will be exported as its own file. File formats are .WAV, .AIFF, FLAC, Bit depth, sample rate, channel order and metadata.

Other Plug-Ins

DTS-X Neural Surround Upmixer converts stereo and surround sound content to 5.1.4, 7.1.4, 7.1.5 and 9.1.4. (See Chapter 8.)

The WAVES MaxxAudio Suite includes extended bass range using psychoacoustics offering better sound reproduction through small speakers, laptops, tablets and portable speakers. Waves has a standalone headtracking controller.

The NuGen Halo Upmix 3D is channel-based output as well as ambisonics. Native upmix to Dolby Atmos 7.1.2 stems and height channel control as well as 1st Order ambisonics. During rendering, the software conforms the mix to the required loudness specification and prepares the content for delivery over a wide array of audio formats from mono to various immersive formats supporting up to 7.1.2. Nugen's software can also down-process audio signals with its Halo Downmix feature that gives the audio mastering process new ranges for downmix coefficients, and a Netflix preset as well.

The Gaudio Spatial Upmix extracts each sound object from the stereo mix and then spatializes the 3D scene on binaural rendering technology adopted from Next Generation Audio standard ISO/IEC 23008-3 MPEG-H.

The Ambisonic toolkit has four different ways to encode mono source: planewave, omni, spreader and diffuser, and two different stereo algorithms.

The Blue Ripple can encode mono sources into a B-Format audio.

The SSA Plug-Ins offers Ambisonic gate/expander, De-essing, rotation, compression, delay and equalizer.

Outside the Box: Black Box Processing

My first experience of a Black box was in 1996, when I used one called the Spatializer. It had eight analog inputs that were controlled by eight joy sticks that could output an expanded stereo – spatialized two channel or a quad output. This device clearly gave the impression of an extended soundfield to the left and right and gave a better impression with simple sources like a single microphone than a group of sounds.

Linear acoustic has designed and built stand-alone boxes for loudness control and management for over a decade. I discussed the new immersive real-time up-processor with Larry Schindel, Senior Product Manager at Linear Acoustic. Linear Acoustic® UPMAX® ISC upmixing (up-processing) can be used to maintain the sound field regardless of the channel configuration of the incoming content. It can also be used creatively to enhance the surround and immersive soundfield.

Audio elements are extracted using frequency domain filtering and time domain amplitude techniques; the LFE is derived from the left, center and right channels without impact to the full range left and right speaker. The surround soundfield can be adjusted via the center channel width control and the surround channel depth controls.

Parameters in the upmixer can be adjusted to help shape the sound for the user's tastes, such as whether the center channel sounds are routed hard center or spread a bit into other channels, or how far back into the surrounds a sound would go to steer upmixed content. The UPMAX ISC can monitor the input signal and auto-detect whether upmixing is needed and native surround content will pass through unprocessed. UPMAX ISC upmix 2, 3, 5.1 and 7.1 to 7.1.4.

Upmixing can be inserted into a mix buss or on the output buss in the OB van or at the network because there will always be a mix of legacy material with mono or stereo sound and

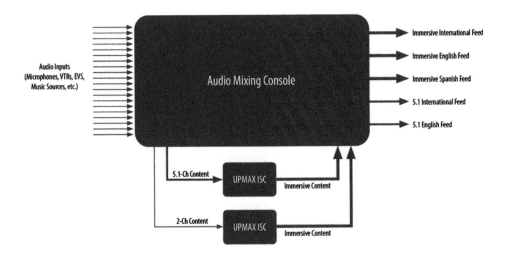

Figure 4.9 Linear Acoustic UPMAX signal flow

it is important to maintain a consistent sound field image all the way through the chain to the consumer/listener. Content passing through in native immersive formats is automatically detected and pass through unprocessed.

UPMAX is a software component included in several Linear Acoustic processors plus is a standalone blackbox for upmixing legacy 2, 3, 5.1 and 7.1 channel audio to 5.1.4 and 7.1.4. Content that passes through the UPMAX that is native immersive will automatically be passed through without processing. UPMAX has been used in live situations for upmixing music, effects and legacy material that is not already immersive. UPMAX I/Os are AES, MADI and SDI.

Illusonics IAP is a real-time immersive sound up-processor. There are features that would appeal to an audiophile although it is an up-processor/sound enhancement device and not a high-end pass through exciter type box to compensate for dull material. It extracts spatial information and creates space around the listener.

If you consider the wide array of inputs, HDMI, Digital Coax, Optical, USB, IAP networks, analog and phono you might think it is a high-end consumer device except for the price tag. HDMI inputs support up to eight channels of 192 kHz and 96 kHz 24 bit audio, the digital coax, optical S/PDIF, the USB port and the UPnP/DLNA network protocol support 96 kHz and 92 kHz 24 bit audio inputs. The outputs are HDMI, balanced XLR and unbalanced RCA. IAP configuration as well as gain, polarity and delay can be applied to input and output channels through your Mac/PC via a USB cable.

There are six adjustment parameters: center, depth, immersion gain, immersion high, immersion size and clarity. Center determines the degree a phantom center is converted to real center. Center increases the sweet spot from everywhere else in the room localizing dialog and soloist in the center of the space. For example, if a stereo signal is selected (2 x mono) which includes dialog, a center channel will be derived. If the HDMI input is accessed the center channel will be directed to the center channel.

Additional surround, height or center information can make depth/immersion more natural, controlling early sound reflections. Depth beyond 50 percent amplifies the rear channels.

Figure 4.10 Illusonics – menu for loudspeakers setups – outputs for 20 positions

Immersion gain is the psychoacoustic sensation of the degree of encircled that a listener perceives. Immersion gain is how strong diffused sound is reproduced. Immersion high – equalization control – brilliance and immersion size – the reverberation time RT60 of the immersion signals. Clarity makes the reproduced sound more dry, reducing the amount of room reverberance and tone controls with bass and treble frequency and gain controls.

Notes

1 Christiaan Huygens. Sciencedirect.com/topics/physics-and-astronomy/Huygens-principle, courses. lumenlearning.com/austincc-physics2/chapter/27-2-huygens-principle, Traite de la Lumiere. Limited John Wiley and Sons 1690,

2 MI. A. Gerzon, "Periphony: With-Height Sound Reproduction," *J. Audio Eng. Soc.*, vol. 21, no. 1, pp. 2–10 (1973 February).

3 Olivieri, Ferdinando, Nils Peters, and Deep Sen. 2019. Review of Scene-Based Audio and Higher Order Ambisonics: A Technology Overview and Application to Next-Generation Audio, vr and 360° Video. *EBU Technical Review.* https://tech.ebu.ch/docs/techreview/trev_2019-Q4_SBA_HOA_Technology_Overview.pdf.

4 D. Sen, N. Peters, M. Kim, and M. Morrell, "Efficient Compression and Transportation of Scene-Based Audio for Television Broadcast," Paper 2-1, (2016 July).

5 Blauert, Jens. 2001. *Spatial Hearing: The Psychophysics of Human Sound Localization.* Cambridge: The MIT Press.

6 H. Haas, "The Influence of a Single Echo on the Audibility of Speech," *J. Audio Eng. Soc.*, vol. 20, no. 2, pp. 146–159 (1972 March).

7 "The Doppler Effect: Christian Doppler Wissensplattform." n.d. Accessed December 16, 2021. www.christian-doppler.net/en/doppler-effect/.

8 McKamey, Timothy. 2013. "Restoration of the Missing Fundamental." Sound Possibilities Forum. September 7, 2013. https://soundpossibilities.net/2013/09/06/restoration-of-the-missing-fundamental/.

5 The Art and Science of Microphones and Other Transducers

Often real-time sonic productions use combinations of organic natural sounds that are specifically captured to enhance and support the production. Sometimes these sounds are altered and manufactured into something that does not resemble the original. For example, the sound of an amplified guitar is sculpted through an amplifier and frequently uses effects for sonic enhancement, but usually the final sound is captured from the amplifier speaker by a microphone. Undeniably each process contributes to a unique tone that a guitarist searches for. Creative sound production can use similar techniques and I am suggesting that sound production for immersive audio is a combination of capture plus creation and modification.

The most basic sound capture is not completely void of acoustic or vibration type properties which can dilute the soundfield with unwanted noise. It is normal to electronically alter or equalize unwanted sonic disturbance from the original sound thus eliminating the problem. Another example is to alter the dynamics of the audio signal because often the energy range of microphones can exceed the electronic capability of the signal chain. However, by dynamically altering the sound characteristics the sound artist is providing an array of options.

The sound of sports has a history of microphone capture but with the advent and growing popularity of eSports there is a purely manufactured aspect of the sound. Even with sound manipulation and manufacturing there commonly was a capture process and then a post-production enhancement treatment. Think of the *Star Wars* light saber where the sound designer took different organic sounds to create a menacing sound that most of us can still hear in our head.

Immersive sound capture is not tied to a specific format or microphone configuration. Immersive sound can be captured with an array or combination of microphones and these transducers can be an assortment of mono, stereo and array type microphones. Your capture goals will direct you to an appropriate selection of microphones. For example, if your goal is to capture the entire soundfield without additional processing, then an ambisonic or an array configuration of microphones would be a good choice. Whereas if your intent is to capture and enhance specific components of the soundfield then individual microphones should be considered to capture the details and then combine and create the soundfield that best characterizes the sonic space. At some point sports audio production will require some specific sound capture from a particular venue no matter how acoustically hostile. Remember, all acoustic capture can be difficult because of the background interference and noise.

Acoustic capture is fundamental to almost all sound production and the sound designer, sound producer and recordist are always trying to acquire specific sonic details within a soundfield as well as the ambient particulars of a soundfield that are required for localization and realization. Most sound is captured by microphones and other vibration-capturing transducers and sound capture is a lot about picking the right microphones and capturing what sounds you want and minimizing those sounds you do not want.

DOI: 10.4324/9781003052876-5

Capturing the Naked Sound: Some Basics

Location acquisition often includes an attempt to capture the micro aspects of a soundfield with minimum acoustic and background information. This can be accomplished by close capture typically using smaller microphones or by distant capture using highly directional microphones. Close capture seems to always be difficult because of size, safety and spook factor of the microphone or operator. These factors account for the reasons why you often have to capture at a distance.

The specific sound(s) are often captured in a single dimension such as mono or an increased dimensional resolution such as stereo, surround and immersive, however each microphone capsule or dimensional increase in resolution can result in electronic noise as well as acoustic background noise. Capturing a single sound element with controlled background for reproduction creates a singular sound object or sound element for immersive sound post-production.

Capturing the macro aspects of the entire soundfield provides the acoustic and sonic information to build an acoustic foundation for reproduction. Larger soundfields usually require a more holistic capture to maintain spatiality and often use spaced pairs of mono microphones or fixed capsule arrays or ambisonic microphones. One spaced pair of microphones creates a basic stereo soundfield, however by adding additional decorrelated capsules you can capture and create a more dimensional 3D sound.

To make a microphone selection, you must first analyze the components of the soundfield and how best to capture it. Sports capture is two separate but equal components of sports-specific and venue-specific (ambiance and atmosphere) and each component generally requires different microphones and sound design.

Focused specific sound capture using passive off–axis sound rejection and filtering is the fundamental principle of removing certain frequencies from the on–axis desired sound. The microphones housing uses openings or ports that allow certain frequencies to reach the backside of the microphone capsule. The tuned opening focuses the front side reception by attenuating frequencies from the sides and rear, thus off–axis sounds are making on–axis frequencies perceptibly concentrated.

Familiar examples of these principles are the cardioid, supercardioid and shotgun microphone designs which are used for focused sound capture. Microphone designs with additional length and a multi-ported design, called the interference tube, are tuned for greater focus and off–axis rejection. Because mono shotgun microphones are designed for off–axis sound rejection, accurate aiming maximizes signal and reduces background noise while sloppy targeting can be disastrous.

Figure 5.1 Student pointing the microphone correctly and not pointing the microphone correctly

Live events are captured in real time using a variety of microphones placed on the sports apparatus, on the field of play, on cameras and around the venue. Through the early 1980s most of the sports and entertainment capture was with mono microphones. Mono Shotgun microphones were adopted from the sound capture methods of the film industry and live sports has remained dependent on distant capture using mono shotgun microphones ever since. Unfortunately, mono shotgun microphones also continue to pick up sounds well beyond the specific sound the shotgun microphone was directed at – plus stationary shotgun microphones have a sweet spot for capture. This is problematic because if you do not continually follow a sound source the results are compromised. The net result of off-axis sound rejection is that the useable sound is only at 100 percent when the sound source is directly in the zone or sweet spot.

Close capture clearly has the benefit of minimizing background noise. By definition, close capture is when the direct sound is dominant over any reflected or off-axis sounds. Close capture for sports and entertainment is rooted with the lapel microphone and as the name denotes it is usually seen as a clip-on microphone on a person's clothing. Lapel, hand and shotgun microphones were typical inventory on the early generation of OB Vans (Outside Broadcast) when rolling the down the road with such a limited inventory, was essential for costs and weight savings. The obvious use for the lapel microphone was for interviews and hands-free situations, but I was introduced to the use of the lapel microphone in a sports application when the lapel microphone was placed in the tennis net.

Since 1996 I have put a pair of miniature lapel microphones in the nets not only of tennis but badminton, volleyball, beach volleyball and table tennis (Ping-Pong). I also began using lapel microphones for "snoop" microphones under player benches, backstage and in the baseball dugout. Because of its size, the lapel microphone has been used on the basketball backboard probably since the first televised game. Size and packaging are an appeal for a microphone's usefulness. The lapel microphone was attractive because of its size and close placement possibilities.

Figure 5.2 Microphone in tennis nets

Production Note

In 1996 I was asked by Bob Dixon, former Sound Designer at NBC Olympics Sports, to add another miniature lapel on the basketball backboard to create a stereo spaced pair under the basket. A stereo pair of miniature microphones in close proximity of the athlete and sports apparatus delivers to the viewer an intimate perspective of the athletes and competition. Thanks Bob Dixon.

In addition to mono shotgun microphones having focused capture characteristics there are computer-controlled microphones that not only focus the capture but also reject the undesirable sounds. This concept is discussed later with ambisonics microphones.

Most of the discussion on microphones has been on mono acquisition, which tends to be a significant aspect of a capture plan and it is easily suggested that much of sports and entertainment sound production is made up of mono sound elements. Capturing the micro-sounds and overlaying them on the appropriate sound foundation is a "layered" approach to sound design and production. Sports and entertainment uses a mono-layered approach to sound capture/production because mono sound elements (objects) can be easily placed anywhere in the soundfield resulting in a composite dimensional soundspace. A layered approach to microphone capture and sound design is dependent on focused acquisition attempting to minimize background noise. Background noise reduction decreases the possibilities of conflicts between different sounds from different acquisitions further down the production process. Layered capture and reproduction gives the sound designer and mixer creative control over the balance and localization of the sound elements, which is the foundation for object-based sound production.

More than Mono

Mono microphones and capsules can also be used in combination for stereo, surround and 3D capture of ambiance and atmosphere. There are formulas like the Hamasaki Square and Cube and variations of the ORTF design which incorporate between four and eight mono capsules respectively, as well as designs by many others that are variations of this technique that seem to capture a 3D soundspace with better clarity and dimension.

Spaced mono pairs of microphones have not only been around for a while but continue to receive the innovation and scrutiny of many audio practitioners and PhDs. Mounting and combining closely spaced transducers is the foundation for many array type microphones capture which have similar applications as other highly directional microphones. Additional benefits result when "beam forming" can be applied. Beam forming is where coincident capsules are combined to form a specific capture pattern. This is discussed further later with ambisonic microphones.

Microphone Innovation

Microphone applications for broadcast and broadband has benefitted from the integration of different technologies as well as advances in engineering, construction, materials and audio adventurism. The lapel microphone was an innovative application of a microphone designed for a different application – this is a natural creative progression. Creative innovation begins by using and testing a variety of different microphone makes and models in unfamiliar and unintended applications.

Figure 5.3 Spaced pair of AT4050 microphones on top of roof in Athens, Greece

---

Production Note

High-quality microphones are nothing new; in 2004 I repurposed five high-quality vocal microphones (AT4050) and put them on the roof of the Olympic stadium in Athens, Greece for eight weeks. These microphones captured the explosions of the pyro-sound with great depth, clarity and believability, and surrounded the listener with the tone of the event. After this experiment, I used the Audio-Technica AT4050 for atmosphere and ambiance at every sporting venue till the stereo version AT4050ST replaced it.

First Degree of Separation: Basic Microphone Capture

Separating sports-specific sound from background noise is challenging and a definite reason for close microphone practices. My education in sports audio capture began in the early 1980s and I was a sponge, soaking up any microphone application that expanded my horizon. Another group of microphones that generally have significant benefits for low-profile placement has been the Boundary, the PZM and the PCC microphone. In 1996 I was introduced to the boundary microphone and used it specifically for close sound capture and for the safety of the athletes in gymnastics and athletics.

The Boundary Effect

The boundary effect is when a sound wave impacts a hard surface and reflects it back, creating a maximum pressure of the sound wave at the surface. Placing a microphone capsule in the maximum pressure zone gives a 6 dB boost in signal. With a properly mounted capsule, the

Figure 5.4 Boundary microphone: 60 degree forward reach from base

direct and slightly delayed reflections combine for a relatively phase-coherent signal with flat frequency response and minimum comb filtering.

The boundary effect is apparent in two configurations – the directional boundary and the reflected boundary pressure zone microphone (PZM).

The directional boundary microphones typically use a 1cm in size or less cardioid or supercardioid polar pattern capsule mounted on a surface. This characteristically gives a 60-degree area of sensitivity above the boundary surface. The direct boundary configuration receives the direct plus the slightly delayed reflected sound where the PZM boundary microphone receives only reflected sounds of the surface.

The PZM – by pointing the diaphragm of the microphone capsule directly at the boundary surface, the reflected sound delay from the surface to the microphone capsule is minimum, resulting in no apparent comb filter interference in the audible frequencies.

Most commercially available PZMs have a relatively small base plate for practical reasons (size) but consider that to maintain the boost in sensitivity across all audible frequencies the PZM should be placed on a large boundary surface such as the lid to a piano. With the additional boundary wall, an additional 6 dB boost can be expected. The frequency peak is determined by the size of the boundary surface, which also effects the color or tone of the microphone.

The PCC Phase Coherent Cardioid is a surface-mounted, half hemispheric supercardioid microphone. The design uses a sub-miniature supercardioid microphone capsule, and since the microphone capsule is placed on a boundary surface the direct and reflected sounds arrive at the diaphragm relatively in-phase. Surface mounting creates a half-supercardioid polar pattern increasing directivity by at least 3 dB. Typically boundary designs improve gain over feedback, reduce unwanted background noise and reject sound from the rear, not to mention are generally small and can be placed closed to the desired sounds.

The directional Boundary, PZM and PCC microphones can be used as spaced pairs to create an image that effectively enhances motion from the movement between two or more microphones. For example, skating microphones can be spaced along the ice and panned between the speakers to hear the skater going by in different speakers or virtual locations in the soundfield. Additionally, the skater's close proximity to the microphones and passing speed creates a slight Doppler effect, contributing to the sense of motion. Even though the direct boundary, PZM and PCC microphones are directional microphones, given their size and the ability to be in close proximity to the desired sound is a significant advantage to clear capture.

Selection Criteria

Finding microphones can be challenging. Microphone testing is difficult in a world where every moment is recorded, broadcast and streamed somewhere. There has been a lot of testing but not much documentation – until now. What are the evaluation criteria which certainly will vary for application? Size, reach, sensitivity, durability and cost are a few evaluation criteria. Size has obviously been a driving factor over the last couple of decades, but microphone technology seems to have advanced using different materials and construction which has resulted in better quality with reduced cost.

Durability is a significant issue for outdoor capture. The moisture, shock, temperatures, much less sound pressure levels are the rigors of sports broadcasting. I have seen a couple of surprises over the years particularly a Shotgun microphone (AT815ST) that was left out in the weather on the infield of the Daytona speedway for one year and still worked. Typically microphones are left in the weather conditions with adequate protection from moisture and wind, and are set up to still function properly. This is covered later in this chapter.

Materials

Conventional microphone capsules have undergone changes in materials and coatings for decades but nothing as radical as the technology in the little microphone in your cell phone. A relatively new technology for microphone construction is called MEMS; micro electric mechanical systems are miniature sensors, actuators and transducers designed for a variety of purposes including for sound. These small transducers convert mechanical energy (vibrations) into an electrical signal using basically the same principle as all microphone design. I was astounded at the quality of this technology and field tested an array of microphones with 19 MEMS capsules. Interestingly this array microphone was designed for music capture, which to me means a higher standard of quality. The capture quality was acceptable but I took the microphone to record the motorsport Drag Racing, and was astonished that the microphone could handle up to about 155 dB SPL before breaking up.[1]

Contact Microphone

As a guitar player I found out early about sound vibrations through solid objects. I got my first guitar, which was acoustic, for Christmas one year and I spent the following weeks figuring out how to amplify it. The contact microphone senses audio vibrations through solid objects and not through air and is generally made of a piezo-electric ceramic glued to a metal alloy disc that is attached to the resonate surface. I found a couple of interesting applications for the contact microphone at the 1996 Olympics, particularly on large wooden surfaces such as on Gymnastic runways and apparatuses, on the Velodrome track and under the weightlifting stage. The success of a contact microphone is dependent on solid contact with the surface. I have had success when I can screw the microphone solidly to the (wooden) surface, but the best application that I have heard was a contact microphone in the ice. See the case study on figure skating in Chapter 10.

The contact microphone was placed in the four corners of the ice rink at figure skating. For most television coverage there are two cameras in the corners, covering the skater at ground/profile level. The skater typically works the routine from the center ice to the corner cameras and as the skater gets closer to the camera the sound mixer can adjust the dynamics/proximity/size of the skate effect moving toward the camera.

Production Note

The ice application works well in moderation and matches a closer action to the camera. I have heard four microphones – one in each corner quadrant and I have heard the addition of ice microphones in the center and middle and prefer to hear the sound come and go in the mix.

Dimensional Capture: 2D/Stereo/Binaural

When possible, stereo microphones should be used to capture sounds to try and reduce any center channel interference – either real or virtual. As mentioned earlier, a pair of spaced mono microphones will capture a dimensional sound. The spaced pair capture method has been used on the hoop in basketball, in the nets at tennis, table tennis, volleyball in goal nets at football, field hockey and handball, as well as with atmosphere capture at most sporting venues. A spaced pair of microphones has proven to deliver a consistent frontal left–right spatial image with a dynamic perspective from the close proximity to the action.

So far we have examined stereo capture using variably spaced pairs of mono microphones. Variable spaced pair means that the microphone capsules are not fixed permanently together and can be moved and tuned. Fixed spaced pairs are closely spaced microphone capsules in a fixed packages. Mid-side, coincident and near coincident microphone are available in a non-adjustable fixed forms such as a stereo shotgun, a stereo boundary or a stereo XY microphone.

Binaural capture is intended to be an authentic capture similarly to the way we hear. Binaural capture (and reproduction) include the effect of the ear – both outside the head and inside the skull – plus any impact from the head and torso.

Stereo capture/reproduction is beneficial to the soundfield for dimensional localization and for reproducing a stable image between the channels. The problem for binaural and stereo capture is you lose the focus of your capture and increase background noise or acoustics.

Figure 5.5 AT4025 XY Stereo microphone

Production Note

A stereo pair can be a correlated pair between any two speakers/channels. For example, I have used the left–left surround, right–right surround channel, left surround–right surround, as well as left and right for imaging and localization of any stereo soundfield. Beginning in 2008, the first full production of 5.1 surround sound across all events and sports I used a left stereo reproduction, front stereo reproduction and right stereo reproduction to reproduce the ambiance and atmosphere. The Game Camera-One (play by play camera) looks down on the field-of-play and this microphone plan matches Camera One perspective/picture and any associated activity in its view such as crowd chanting or the wave.

Binaural Recording

There seems to be little binaural capture in sports and entertainment because of cumbersome, limited and expensive options for properly mounting the microphone. In 2017 Sennheiser released an interesting binaural configuration that mounts a medium quality transducer/microphone in each of the earbuds. The earbud fits like any earbud but the microphones are mounted where your ear canal is, in a true binaural location with the capture influenced by the wearer's head, torso and pinna – outer ear.

Immersive sound production still requires acquisition basics with transducers, just like any professional sound production.

3D Microphones

As the sound designer (sound recordist) moves from capturing only in the horizontal plane, not only does the channel count increase but also the complexity of the sound scene increases, along with additional complication for implementation. For example, soundfield complexity could be described as a dense soundscape with many sound sources from all directions such as a large stadium or city setting. Listen to Dr. Lee's City Soundscapes. The most basic of dimensional capture can be accomplished with a 1st Order ambisonic microphone but because of the low resolution there is very little detail. While fixed arrays as proposed by Dr. Lee and Dr. Helmut Wittek provide good signal separation, diffuse field correlation and a very stable soundfield in the listening area, the impact of a complicated setup cannot be underestimated – stands, cables, weather protection plus cables which require time to prepare.

What is the perspective of height? Reproducing sound elements that are familiar and are usually above the listener clearly delivers the perspective of height. This is localization and realization. For example, when the listener hears an airplane in mono, the localization is fixed but the realistic aspect of the sound is from moving created by volume and acoustic manipulation. When the sound designer reproduced the airplane in stereo there was an increase in believability because the airplane could move in the soundfield – left, right, front or back.

Sound above the listener is critical for the reproduction of immersive sound. Sound from above is known as height sound and is represented graphically as the vertical or Z axis. With the 3rd dimension, the vertical Z axis was added so the viewer has full belief that the airplane is moving above the listener because the listener hears the airplane moving above.

Expected sound above the listener creates the illusion of height, as does reflected sound. Reflected sound gives a dimension of spaciousness because the listener hears a direct sound and a combination of reflected sounds some period of time later. The extreme of audible reflections is reverberation, which is extremely diffused reflected sound such as in a cave or cathedral. Is spacious sound immersive sound? The first question is what you are trying to capture and why. Secondly, what is the condition and complexity of the soundfield? Let's examine an airplane scene. The perspective of the sound is from ground level and the goal is to make as clean a capture as possible from the ground perspective. The condition and complexity of the capture depends on the level of background noise and ambiance. If the sound capture of the airplane is for multiple uses, then the neutrality of the background is essential and the appropriate method of capture may be a mono shotgun microphone with an operator tracking the airplane. This method gives a consistent capture of the airplane within the sensitivity range of the microphone and can be believably layered over virtually any soundscape.

In our previous airplane example we captured the detail of the engine noise but we still should spatially capture the soundfield and ambience separately. With a spaced array capture method the airplane, spatial dynamics and soundfield are all captured at the same time. The location of the airplane is captured as the movement changes the sound intensity from one capsule to another while the surrounding sound elements stay relationally and dynamically the same to each other.

Capturing a detailed dimensional soundfield is a difficult undertaking. Mobility, rigging and stability, along with setup time, are a major consideration for field recording. I was the sound designer for 28 years at the Olympics and I was often criticized for having over-the-top sound designs that required extra and extensive setups. I would agree but the knowledge gained at the Olympics, where there is extra setup time and extra equipment, trickles through the audio community via the international crew that works on the Olympics.

Variable Spaced Arrays vs. Fixed Spaced Arrays

Fixed arrays are microphone capsules that are physically connected through a fixed harness or housing. The capsules in these types of arrays are often physically very close together and cannot be adjusted. Common is the concept of a single point capture where the microphone capsules are tightly fitted on a sphere. Often these types of fixed close microphone capsules are associated with ambisonics and spherical capture. See Chapter 4, Ambisonics.

Variable spaced arrays usually have to be assembled and usually there is a specific formula that determines the separation of the capsules. Separation of the capsules can give more of an illusion of spaciousness and I have found that widely spaced pairs of microphones work well in large sports venues.

Sports venues are often a large and diluted soundfield, which is difficult to capture from a holistic approach. I have had success by widely spacing pairs of microphones above and directly in front of spectators, avoiding any direct sound from a PA speaker but the separation of the microphones in my sound design can be complicated by practicality, making equidistance placement not possible. I break the venue into quadrants, and place the microphones as equally as possible in the venue but the distances can be upwards of 50 meters. This works because there is still sound correlation as the sound generated in a sports venue is generally homogenous and uniform throughout the space. This has been confirmed from measurements in several different Olympic venues in different Olympic years.

I listened to Dr. Lee's sonic landscapes at AES 2019 and discussed with him his principles of what he calls Equal Segment Microphone Array 3D. I liked his concepts and believe that there is a similarity in his research and my previous work.

Dr. Lee has recorded sonic landscapes using his variation of spaced microphones named Equal Segment Microphone Array (ESMA). These principles capture the entire 360-degree

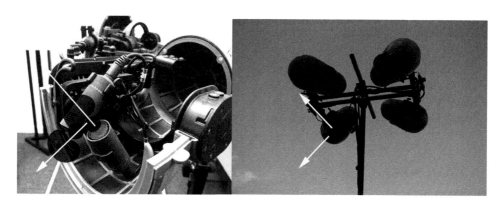

Figure 5.6 ESMA 3D microphone array

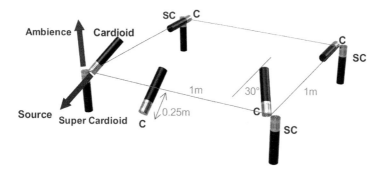

Figure 5.7 PCMA-3D Array

horizontal soundspace and Dr. Lee has proposed a vertical capture scheme that is compact and effective – Equal Segment Microphone Array 3D.

I was intrigued that Dr. Lee suggested that each corner of his array was a vertical coincident pair/vertical mid-side pair and could be decoded into a downward-facing and upward-facing virtual microphone. This concept may be useful for sports because virtual phantom images can be perceived between the loudspeaker layers, giving greater creativity and flexibility to the immersive sound designer.

Dr. Lee's ESMA-3D design captures all relevant information around and above the listener. For example, the 3D microphone array captures the airplane, airplane movement and all ambient sound around the capture point. The ESMA-3D captures a 3D acoustic fingerprint of a space and can be a valuable tool for the holistic spatial capture of ambiances and atmospheres. Dr. Lee's ESMA-3D Perspective Control Microphone Array uses wider spaced pairs of microphones to produce 11.0 (7+4).

I have tested stereo microphones that are oriented vertically instead of horizontally – top to bottom instead of left to right – to capture more present sounds from above. This is similar to Dr. Lee's design; he said that his PCMA-3D design is exclusively a vertical coincident pair and that he normally use supercardioids facing upwards for the upper layer, and cardioids facing towards the source for the middle layer.

Dr. Lee constructed a quadraphonic layer at ear height, augmented with another quadrophonic layer elevated at 30 to 45 degrees, cardioid or supercardioid microphones facing directly upwards. According to Dr. Lee the height aspect can be achieved with microphones mounted like his ESMA-3D design where each corner of the square has a pair of mono microphones in near coincidence or mid-side configuration.

I found Dr. Lee's vertical spacing proposal interesting particularly since his research concluded "that vertical spacing between the main and height microphones of a main microphone array had no significant effect on the perceived spatial impression".

For sports sound this concept can be a relevant and supporting principle for the widely spaced pairs that I have used for 20 years at the Olympics. Clearly there is a relationship between the separation of the capsules and the uniform capture of a soundfield but I wonder if there is an optimum spacing for the size of the soundfield?

ORTF-3D Array

Dr. Helmut Wittek from Schoeps Microphones has spent decades recording and studying a wide range of sound images, including music and sports, as well as having significant involvement in the sound design of multiple World Cup Football Championships. Dr. Wittek has published and field tested several original variations of the ORTF array principles for immersive sound capture. The ORTF-3D uses 8 supercardiod microphones in two concentric planes. There are four microphones on each level, forming rectangles with 10 and 20 cm sides. The vertical imaging is produced by angling the microphones into 90-degree X/Y pairs. The top level of microphones are pointed up while the lower level are pointed down, which easily creates separation in the vertical plane. Dr. Wittek told me that "such a two channel coincident arrangement works well because of the highly directional supercardioid capsules and that the image quality and the diffuse-field decorrelation are both good and that an even better decorrelation could be reached by spacing the microphones further apart."

The four lower microphone signals are routed discretely to the L, R, LS and RS channels, while the four upper microphone channels are directly routed to the Lh, Rh, LSh and RSh channels. These channels represent directional audio information and do not have to be decoded from one format to another like ambisonic microphones, which makes it possible to use this microphone plan live with no format coding latency.[2]

Fixed arrays can only capture a zone of sound because of the limitation in reach of the microphone capsules and background noise. Reach is a product of microphone focus (supercardioid) and the background interference with the desirable sounds.

Often when considering a sound design for sports the question is what are the relevant height sounds? Natural height spatialization for sports seems to be more atmosphere

Figure 5.8 The ORTF-3D: 90 degrees of separation

or ambiance, which is a broader approach to the soundfield as opposed to a more compact localized approach. Significantly, sports and entertainment productions are reluctant to allow microphones to be visible and both arrays are noticeably large and would be restricted. Good news – both arrays are well suited for *anchor microphones* to provide a stable dimensional soundfield, relevant to the venue/space for spot microphones to be added. In a creative sound design, height details can vary from expansive to compact depending on the event or sport. Football is a focused sport and the objective is to follow the ball and the height sound will inevitably be atmosphere and PA.

Gymnastics is a group of individual sports under the umbrella of men's and women's disciplines that are presented simultaneously till final events. Gymnastics production is usually visualized from a close camera perspective and needs a very specific sound design that is closely focused on the apparatus and athlete.

There are detailed and specific reasons for the spacing between microphone capsule as described by Dr. Lee and Dr. Wittek, however I am presenting some practical reasons for why I designed variations on the theme which delivers extreme detail and a defined 360 type spatialization.

Inverted Equal Segment Microphone Array

At the Olympics I have used the apparatus, in this case the parallel bars – see drawing as the mounting points for microphones for several decades. With surround sound the capture point is in the middle of the vertical bars on each corner of the parallel bars. This delivers the ear-level sound quadrangle and with the addition of a downward focused quadrangle we now provide a 360-degree perspective of the athlete in competition.

The parallel bars are the base layer for a variable and inverted equal segment microphone array. Variable means the mounting pattern on the bars is a rectangle not an equal-sided square and inverted because the microphone capsules are pointed down and not up as opposed to the equal segment microphone array.

When building a variable and inverted equal segment microphone array – 3D on site, use miniature cardioid capsules for the lower four microphones and a quadrangle of supercardioid or shotgun microphones mounted on the lighting truss pointing down to be reproduced in the height speakers.

This sound production for the parallel bars is a complete 4.0.4 immersive soundfield and can be seamlessly cut into the integrated production, play as a simultaneous and continuous sub-production or be recorded for delayed play out with eight channels of audio. Global presentation of multi-discipline sports such as gymnastics and athletics for international distribution would typically have individual feeds for bars, rings, jumps and floor exercise, plus an integrated feed. During the early days of the competition there is continuous and simultaneous competition and compact coverage of the individual disciples attempts to minimize adjacent noise from other competitions.

The 4.0.4 – concentric – quad coverage is very localized and specific to one apparatus. This microphone plan accounts for no crowd capture except over reach of the microphone beyond the specific apparatus sounds. In addition to this focus plan, there should be additional coverage of the immediate audience that will be reacting to the competition in front of them. Localized capture of the ambiance and atmosphere is an ideal application for the equal segment microphone array – either the ORTF-3D or the ESMA-3D or some variation as discussed earlier.

This array design is not intended to be summed to surround or stereo, which should be rendered separately.

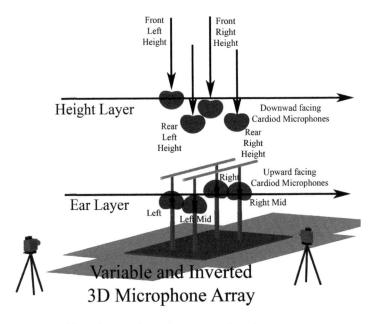

Figure 5.9 Variable and inverted equal segment microphone array

Figure 5.10 Miniature cardioid microphone being mounted at the flex joint on the parallel bars

Gymnastics is an example where there are specific zones in the venue where the audience may react to what is directly in front of them, whereas in football, the audience is watching the same action but one area (home team) is not reacting the same as another area such as the visitors.

At events where the audience is more disciplined, for example at tennis events there is a more homogenous dynamic nature of the atmosphere sound across the entire venue because there is less of a tribal-seating effect.

The major problem with microphone placement is they are often in the way visually or in the way for an overhead aerial cameras position, which is more and more common.

An ORTF-3D or the ESMA-3D above and in the center of the Arena could ideally capture the acoustic fingerprint of that venue. In a tennis sporting venue, a center microphone cluster could easily capture all levels of crowd seating and the microphone cluster could be aimed downward at the court. The concentric quadrangle aimed at the crowd and at the activity (parallel bar or tennis court) would be an optimum combination of microphones and mounting for a 4.4 foundation.

Production Note

The possibility of a single cluster above the FOP is unlikely in many if not most situations and this is exactly why our variable and inverted equal segment microphone array – 3D is a valid principle.

A concentric quadrangle approach to microphone configuration and capture appears to deliver predictable results. This chapter has looked at variations including traditional equal segment microphone array variable as well as 3D and inverted, and all appear to be applicable and relevant to dimensional sound design.

A simple principle is that variably spaced microphone clusters, in homogenous soundspace such as outdoor stadiums and large indoor venues, work well to capture the spaciousness of the soundfield. Perception of spaciousness is dependent on the capsule configuration and spacing.

Sidebar: Sports Sound Design

The man in the stand's philosophy is from the perspective of fan in the middle of the stands. Note the front to back perspective of the listener from diffused to sharp. As you imagine the sound above you, from ear level to above your head is from sharp at ear level to diffused above and around the listener's head.

Layers of spaced pairs of microphones are specifically placed (tuned) to capture a tone or layer for the above stratus as well as at ear level. Remember, you do not have to capture immersive sound to create immersive sound.

Arrays, Tree and Ambisonics Microphones

All of the previously discussed microphone configurations capture the soundfield in a manner that makes it possible for a realistic dimensional soundfield reproduction.

Closely spaced microphone clusters and widely spaced microphone combinations all can create immersive sound, however the difference in capture methods and configuration results in differences in the accuracy and resolution of the capture.

All microphone capture is problematic because of the difficulty of a clean capture without background noise which generally makes most microphone capture somewhat localized. Bottom line, microphone placement will always be a problem because of the fact that a single microphone or single closely spaced microphone clusters can only capture a slice of the acoustic space which clearly results in a low accuracy and resolution beyond the immediate sensitivity. I agree that a single microphone position is capable of capturing all necessary and proper perspectives to the sides, behind and above the listener, but this capture may be only very immediate to the vicinity of the microphone configuration.

Sports microphone capture has been often about how to amplify the reach. The cardioid, supercardioid, shotgun microphone and the parabolic microphone were all designed for focused capture and transducer engineers have carried forward the quest using multiple capsules and computer control.

Advanced concepts of focused capture have to do with forming capture areas, known as beams or sweet spots, from what are known as array and spherical microphones.

An ambisonic microphone uses multiple closely spaced microphone capsules that work together to capture an ambisonic soundfield. Array microphones come in a variety of forms, but all share in the ability to adjust patterns and directivity as a result of computer modeling. Computer-controlled microphone modeling can be accomplished with as few as three (two) capsules and I know of a 31 microphone horizontal array and a 64 capsule spherical array. More capsules and the combinations of microphone capsules in coincidence or as near coincident as possible give the possibilities for better directional resolution and directional rejection.

Similar in principle to the ORTF-3D or the ESMA-3D, the ambisonic microphone captures the entire audio soundfield (scene) of the sound coming at the microphone from all directions – front, sides, back and above. Similarly, the microphone is the center spot of the capture similar to the way we hear and when the soundfield is reproduced, the capture is the listener's POV and sweet spot.

The concept and math of ambisonic capture came out of the UK from Dr. Gerzon in the early 1970s as he introduced 1st Order Ambisonics which captures directional information from all 360 degrees, similarly to the ORTF-3D or the ESMA-3D but with a big difference: Dr. Gerzon places four microphone capsules as closely as possible to the axis of a sphere – full sphere can record equally (calibration) from above and below as well as all horizontal directions. 1st Order Ambisonics is an extension of mid-side stereo (see Chapter 4).

The four microphone capsules capture a three-dimensional soundfield to four channels – W, X, Y and Z. W is the omni directional information, X direction, Y direction and Z direction. The mono audio signals are captured at the prescribed horizontal angle (X and Y) – azimuth and vertical angle (Z) – elevation.

1st Order – four capsules – 1 × omni directional and 3 × figure-of-eight microphone bidirectional capsules, ideally coincident at a point in space. In practice, it is impossible to fit all three microphones in a single point and a practical design places four cardioid or subcardioid capsules in the vertices of a tetrahedron arrangement. The capsules point Left-Front Up, Right Front Down, Left Back Down and Right Back Up.

Ambisonics seems to be a natural evolution of the Blumlein Crossed figure of eights and MS microphone. The Sennheiser Ambio microphone uses the same tetrahedral arrangement as the original Soundfield design and captures the A-format four channel output that has to be converted to the ambisonics B-format. Note – the tetrahedral design is in the public domain and has also been used by other manufacturers.

The raw recording of the four microphone capsules is Native or A Format. One channel carries the amplitude signal while the other channels represent the directionality through phase relationships which requires combining the channels.

The four capsules are Lb, Lf, Rf and Rb. The A Format signals have to be converted to ambisonic channels – W, X, Y and Z known as B Format. The B Format matrixes the raw audio, which is considered to be 360 sound by combining the omni and the three bidirectional channels. Mixing to B-format combines the channels as follows: W = FLU + BLD + BRU / X = FLU + FRD – BLD – BRU / Y = FLU – FRD +BLD – BRU / = FLU – FRD – BLD + BRU. The B-Format supports 16 channels or 3rd Order ambisonics.[3]

Ambeo A-B Format converter is a plug-in that not only converts but controls the audio signals with filters, orientation control, output format and adjustment controls. The microphone can be directed upright, upside down or forward – endfire. The default position is upright – there is a forward mark on the casing.

Output format determines the order of the four channels. With FuMa the channels are ordered as W-X-Y-Z and with ambiX the channel order W-Y-Z-X and are obviously not compatible. Additional plug-ins have been jointly developed like the dearVR MICRO and ORBIT and FOCUS. dearVR MICRO is a set of HRTF and binauralization filters based on the Neumann KU100 dummy head. dearVR ORBIT allows the user to control the acoustics of a virtual room by adjusting the size and reflection level. Additionally you can alter the timbre of an acoustic environment by changing wall, ceiling and floor covering. The FOCUS parameter lets the user balance between the spatial perception and clarity.

After being converted to ambisonics, the four-capsule configuration can be combined in many desired polar patterns, such as adjustable mono, stereo or 5.1, as well as used for binaural 3D reproduction essential to head-tracking and VR.

Live Workflow

The Ambio ambisonic microphone must be converted from the raw A-format to the useable B-format which is hosted on a standalone computer. Given any conversion from A-format to B-format and the output from the computer to the output to get to the mixing console there will be a noticeable amount of latency, however the use of this microphone to capture ambiance and atmosphere where visible lip-sync would not be apparent, as in most sports venues, would be acceptable.

The Soundfield 1st Order ambisonics microphone is a broadcast microphone that has virtually no latency and has been used extensively in the UK on football and has been the foundation of BBC and Sky football coverage for decades. Ambisonics has since evolved with the development of Higher Order Ambisonics (HOA), which is a multipolar expansion of the soundfield capture around the singular, spherical capture point by simply adding more symmetrically directional captures at the sphere.[4]

HOA increases the resolution of the soundfield capture by increasing the number of microphone capsules/channels. As you add microphone capsules, the order of ambisonic increases along with additional amounts of detail. 1st Order – 4 capsules, 2nd Order – 9 capsules, 3rd Order – 16 capsules and 4th Order – 25 capsules or more. Note the first 4 channels is always the same for 1st, 2nd, 3rd all higher orders which means you can examine the differences in resolution between orders by subtracting or adding back channels.

HOA is scalable. For example, 1st Order Ambisonics is four channels and 2nd Order grows to 9 capsules/channels, 3rd Order is 16, 4th Order is 25 and so on. The formula is: Order+ 1Squared. For example, 5th Order is 5+1=6Squared=36 Channels/Capsules. 2nd Order ambisonics is made up of the 0th Order (Omni) + 1st Order signals and the additional 2nd Order signals. 2nd = Oth + 1st + 2nd Order information. The 2nd Order ambisonic microphone outputs standard 9 channel B-format, however as you increase resolution you increase processing latency.

Core Sound's Octomic (www.core-sound.com/OctoMic/) is a 2nd Order ambisonic microphone that outputs audio signals and does not need a computer to record but as with all ambisonics it must be converted to 16 channel B-Format through some app. 1st Order ambisonics is really only able to accurately reproduce 7.1 because it has limited height information. 2nd Order ambisonics seems to bring a moderate increase in height information.

The ZYLIA ZM-1 is a 3rd Order ambisonic microphone array that uses 19 omni directional MEMS capsules. This technology is interesting because the microphone is completely digital without analog capsules, preamplifers and A to D converters which can add floor noise. MEMS – Micro Electro Mechanical Systems is a silicon technology that has small fabrication

Figure 5.11 Octomic 2nd Order ambisonics microphone can be plugged directly into a recorder

form, fabrication repeatability, stability in extreme temperatures and low power consumption, making for more consistency between capsules. This microphone uses a USB interface to power the electronics and capsules and currently needs to be recorded to generate a file for post-production, although there is the possibility for real-time monitoring of combinations of capsules, but you cannot output audio channels discretely or monitor ambisonics.

With the latest version of the ZYLIA ZM-1S another significant addition was a timecode input to facilitate synchronization with other microphones and video footage. The latest version also supports the emerging technology 6 Degrees of Freedom Navigable Audio. I have demoed 6 Degrees, which seems to work very well, and I expect this innovative technology will find a place alongside other new generation realities such as AR, VR, MR and ER.

Although Zylia began with a vision to simplify recording music, I have recorded college football and basketball, a steam tractor parade, a classical Christmas recital, my Farmall Tractor plowing a field, 4th of July fireworks, and Drag Racing 50 feet away from the track (see Chapter 10), plus the inside perspective of the audio control room of that drag-racing event with the ZYLIA microphone ZM-1. I think that because ZYLIA begot as a musical-orientated company they paid a lot of attention to tone and timbre; as a musician I can tell you fidelity and sonics are first.

The ZYLIA engineers have created some useful tools in their apps. The software is good at isolating the sound source providing close to full separation of sound sources for later rebalancing. Advance beam forming software provides spatial filtering and algorithms to adjust amplitude and phase to provide a wide range of polar patterns.

This is similar to all ambisonics microphone, however as you move to 3rd Order ambisonics from 2nd Order the microphone seems to provides a significantly greater level of detail than lower orders. As a recording transducer 24bit, 48 kHz, 128 dBSP with only 1 percent distortion with timecode is appealing for broadcast application as well as the new realities.

My biggest surprise was when I took the microphone to the drag racing track. I did not measure the sound although I was told it was around 165 dBSPL at trackside. I was about 50 feet away with a laptop and the ZYLIA ZM-1 microphone on a fishpole with triple ear protection. I had experienced the intensity of the sound at around 100 feet but decided to see what would happen at 50 feet. Drag racing is not a consistent soundfield but a burst of sound that does not last more than about five seconds. The car took off and the sound and sound pressure was so intense that it physically moved me. I padded down the microphone input as much as possible as I pushed record. The computer screen shut down immediately and I was not sure what I had till I went to my car to see what had happened to my computer and the ZYLIA microphone. I was surprised that the microphone holdup and the computer screen went black but continued to record with a little digital saturation at only one brief moment. I was surprised and so were the folks at ZYLIA. The ZYLIA Control Panel has a single control for 70dB of master gain.

Once you start a session ZYLIA Studio has a three-step calibration process. First, select a representative sound source. A wide range of musical instruments represents a wide range of source levels and frequency content. I have used "other", "electric guitar" and "drum" – obviously loud settings. Next you record an eight-second passage of the sound source to further tweak my non-musical instrument pre-calibrated selection and any fine adjustments can still be done through the ZYLIA Control Panel.

The microphone's pre-calibration setting and possibilities optimize the dynamic range of that particular recording and can be calibrated for a quieter session such as an acoustic guitar, voice and violin or the extremes like drag racing.

Next a screen comes up to check if the position of the microphone is correct and then you are ready to record.

On the July 4, 2019 I recorded a parade of steam tractors and later that day a pyrotechnics show. Both recordings were very dynamic and directional. The steam tractors were particularly loud with their whistle blows and there was one tractor with an out-of-tune calliope. I recorded the parade of about 20 tractors from 25 feet away on a slight hill behind the street crowd. The 3rd Order ZYLIA had no problem with the dynamic range and clearly reproduced the hemispheric image in front and above me with clarity.

I recorded a plowing sequence with my Farmall Tractor where my wife and son drove within 1 meter of the microphone and made wide sweeps up to 30 meters away. The recording had a wide dynamic range because of the wide path of the tractor passing by the microphone where the microphone spatially captured the tractor from every direction and distance.

I recorded a college basketball and football game at Middle Tennessee State just south of Nashville with the Ambeo, ZYLIA, Eigenmike® array and four uniformly spaced stereo microphones. My evaluation of these side-by-side captures and a classical Christmas recital with the Zylia 3rd Order, a 4th Order ambisonic and constructed arrays appear later.

My experience with ambisonics is that even with a higher order ambisonics capture, the soundfield is so complex that there is usually too much frequency and volume masking for a detailed capture, which results in additional spot microphones. Note that spot microphones and other mono and stereo sound objects do not capture a 3D soundfield and do not have the spatial characteristics of an ambisonics microphone. The results are that when combining ambisonics material and spot microphones your 360-degree soundfield can be rotated but the objects will have a fixed location.

Figure 5.12 ZYLIA control panels

The timecode input for the ZYLIA ZM-1S cannot be underestimated. I have made many side-by-side recording using a manual clap board for sync, which works but can be challenging when synchronizing microphones that are spaced far away and/or in a space with lots of reverberation and delays.

In a short period of time ZYLIA has made major advancements in their microphone and software, particularly with the addition of timecode for synchronization and the very innovative software to support the concept of 6DoF – six degrees of freedom and volumetric 3D audio experiences. ZYLIA's 6DoF navigable software lets you place multiple Zylia microphones in a soundfield and effectively move and rotate through a soundscene and hear everything dimensionally through the listener's perspective.

Production Note

I interviewed Josh Daniels the soundmixer from drag racing on a lapel microphone and had the ZYLIA in the control room for an ambisonic bed. I added my outside recording to the mix and the immersive feel was incredible. This could be a big plus for radio drama.

There clearly is a noticeable difference in the detail of the soundfield and better control of the beams as the ambisonic order increases. The well-known 4th Order Ambisonics Eigenmike array has become a reference standard for ambisonic capture and research.

The Eigenmike array is a 4th Order ambisonics microphone that uses 32 analog electret dynamic capsules that, when combined create Eigenbeams, also known as HOA coefficients that capture the soundfield as well when combined steer multiple simultaneous beam patterns.

Dr. Gary Elko, creator of the Eigenmike array, told me that:

> The spatial polynomials that the array is resolving for a specific order is (order+1)^2 however that is a lower bound. We could have gone higher with the Eigenmike Array em32, but we could not have gone to the next order. We did oversample the surface, but the geometry of the microphones was good as it allowed for use to combine the microphones with equal weight in the gaussian quadrature that is part to how the sphere is used to decompose the spherical harmonic modes. The 32 microphone locations essentially satisfy the orthonormality criterion in that the weights for the spherical harmonics are such that if multiplied by other sampled spherical harmonics would yield a discrete integral of 0 thus fulfilling the discrete orthonormality criterion that is desired for the spherical harmonic decomposition. This behavior has nice implementation aspects and that is why we used that. There are an infinite number of geometries and there are some good ones and this was one of them and FireWire supported 32 microphone channels.

The new Eigenmike Array em64 does use quadrature weights but they are all close to 1. Oversampling the sphere also pushes up the frequency where spatial aliasing starts and that is around 8.5 kHz on the em32 and 12 kHz for the em64 (same diameter as em32 but with 64 mics).

The new em64 uses power over ethernet and Audinate's Dante protocol and you can combine up to eight on one ethernet network. The Eigenmike array can run in excess of 100m on CAT5 or better cables to the microphone interface box that converts the Eigenmike to a Firewire audio stream to the PC that controls the 32 channels in EigenStudio. The

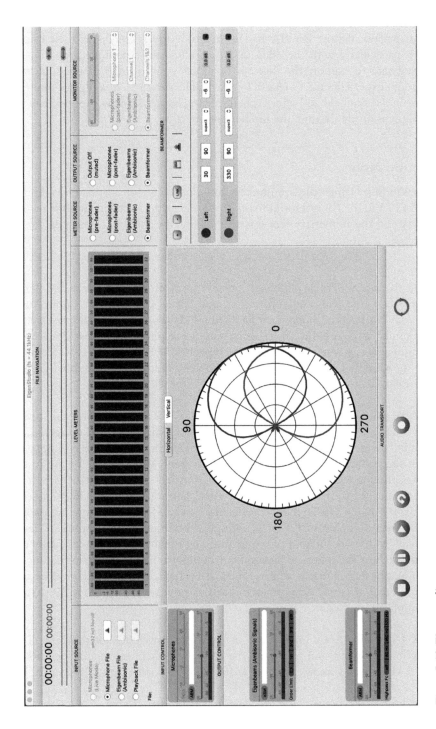

Figure 5.13 Eigen studio

Figure 5.14 Eigenmike array microphone capsule configurations

EigenStudio applications allow for real-time beam steering and can perform real time spatial analysis

With every higher order ambisonics microphone, as the resolution increases so does the processing time because of the A-format to B-format conversion. Latency further complicates the use of ambisonics in live situations because of latency.[5]

The Sennheiser Microphone Array (prototype) has 31 shotgun type microphones and is differentiated by its design to capture primarily 360 degrees, but only in a single horizontal axis. These microphones are oriented facing forward from the mounting sphere with all microphones aligned on a single horizontal plane. Sennheiser says that this type of array cluster can process up to four focused sound beams plus can independently steer each beam's focus. For example, the microphone can optimize the capture of the sound of the ball, the coach or any sound source in the horizontal soundspace and output each beam separately. But this array cannot output an ambisonic soundfield because this array captures no discrete height information. Of concern with this microphone is the form factor. The current Sennheiser prototype uses 31 short shotgun type microphones and the form factor is large, but ultimately they expect the final product to be half the size while still using 31 directional microphones. The shotgun array focuses the capture on the horizontal axis while ambisonic capture attempts to capture and reproduce the entire soundfield as possible.

The original Soundfield microphone ST450 Mkii 1st Order ambisonic microphone requires the Soundfield SPS422B processor box and was designed for broadcasters.

Figure 5.15 Sennheiser microphone array (prototype) has 31 shotgun type microphones

Benefits of Higher Order Ambisonics (HOA)

Higher Order Ambisonics (HOA) can enlarge the quality and size of the reconstructed soundfield, and expanding the sweet spot for broadcast and broadband is a desirable result. Additionally, ambisonic capture is beneficial because it provides head tracking and matches the panning/head panning of the consumer.

Beam Forming

Microphone modeling – once you go to 2nd Order ambisonics you can effectively and accurately model most microphone polar patterns. Beam forming – array microphones are capable of beam forming resulting in focused accumulation of sound through microphone modeling and computer manipulation. Beam forming is an active approach using multiple fixed microphone capsules and creating a variety of patterns or directions using the fixed capsules to optimize the capture of the desired objects such as the ball, the coach or the player. Not only does beam forming benefit from focused capture, but also rejection of unwanted off-axis sounds.

A final benefit with beam forming microphones is the ability to steer the microphone beam dynamically and in real time using positional information from video systems like the Chyron Hego. The Chyron Hego uses cameras to track objects on the field-of-play and creates positional information of where an athlete, ball, coaches or even a car is located. This data can be used to steer an audio beam to optimally capture and even follow the desired sound. Question: Is Beam forming still ambisonics?

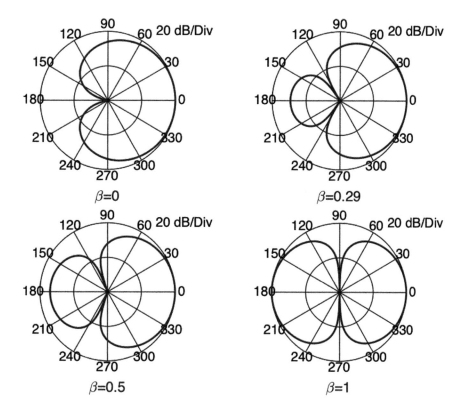

Figure 5.16 Beam forming polar patterns

Ultimately, chasing the sports sound is a fatiguing and a non-gratifying experience while beam tracking is a tool to support the innovative audio mixer. Novel microphone application and design are essential to advancing the art and science of sound and are the responsibility of both the users and the dreamers. The designers and manufacturers of audio equipment want good input and sound advice – and it is always a good time to participate.

Sound Design for Immersive Sound

You do not need array microphones to track a sound or ambisonic microphones to create immersive sound, but these are powerful tools to be considered for advanced audio production. Immersive sound design and production seems to be a combination of holistic and specific/object audio production and without doubt immersive sound is the final frontier in dimensional sound capture and production.

There are obvious benefits to 3D capture with various combinations of spaced microphones and capsules. However as mentioned earlier, sports capture will require multiple acquisition points because of a high degree of noise and physically large audience space.

There are advantages for 1st Order ambisonic microphones, such as the Sennheiser Ambio. They are effective and economical packages that can deliver a stable/cohesive immersive sound bed and are a small package with virtually imperceptible latency.

Moreover, the price of these microphones allows for ambiance and atmosphere capture from multiple capture zones. Multiple capture points deliver an acoustical representation of the venue space with on–axis presence and natural diffusion from the venue architecture.

Production Note

When capturing 3D sound for VR/AR/MR the microphone and the camera share the same reality position and space and must be placed as close together as possible. I have seen sound designs that place a multi-capsule microphone array center on the near side of the field usually mounted over the audience. This position, along with some far-field spot microphones, is intended to be the basic sonic foundation of the immersive mix. This sound capture tends to be heavily weighted to sound capture of the near-side audience where the intention is to provide a wide spacious ambiance that matches the play-by-play camera looking down from mid-near center. This camera spends a lot of time on the air because you can see the entire field of play and three-quarters of the audience.

Intuitively it seems like a good idea to anchor the generic atmosphere base to the wide play-by-play camera one position. It is a good start, however without similar sounding capture points spaced around the venue the mixer/sound design has a limited sound perspective of the venue and they may be stuck with unwanted noises.

Variable Spaced Pairs

Selecting microphones and placement to cover/capture the left, far and right side of the arena creates a widely decorrelated design that reasonably and spaciously covers the sonic finger-print of the stadium.

I currently use stereo microphones in these positions and pan the microphones as follows: left stereo microphones is panned left surround and left mid – the left far microphone is panned left mid and left front, the right far microphone is panned right and mid right and the right stereo microphone is panned right mid and right surround. I call this design "variable spaced pairs" because the sound designer/mixer can vary the space between the capsules to tune-in desired sounds and tune-out unwanted noise – particularly PA. I like to have my atmosphere/ambiance microphones in front of the crowd and preferably on stands so the microphones can be moved when necessary to tune the venue for a more constructive powerful ambiance and atmosphere sound capture.

Finding acceptable places for the microphones and stands can be challenging when you have to consider fans line-of-sight and PA dispersion. The microphones may have to be in places/spaces that are not non-symmetrical to each other, however precise physical placement of the microphone is not as critical with large spaces where acoustics create a fairly homo-genous sound throughout the venue. Rigging the microphones over an audience can be very effective, however moving the microphone during live action can be problematic and the sound mixer may be stuck with some undesirable noises like a vuvuzela. See Chapter 10.

I began using the left-mid, far left, far center, far right and right-mid microphone capture scheme for atmosphere using mono microphones and about a decade later I started using the Audio Technica AT4050ST stereo microphones. Each microphone position is a stereo field and should be panned as previously described, however with stereo microphones you can capture a wider spread and localization. For example, the left mid stereo microphone would be panned L–R but that translates to a more wrap around feel to the left rear and right rear.

I have tested stereo microphones that are oriented vertically instead of horizontally – top to bottom instead of left to right – to capture more present sounds from above. This is similar to Dr. Lee's design who said that his PCMA-3D design is like a vertical mid–side pair and my

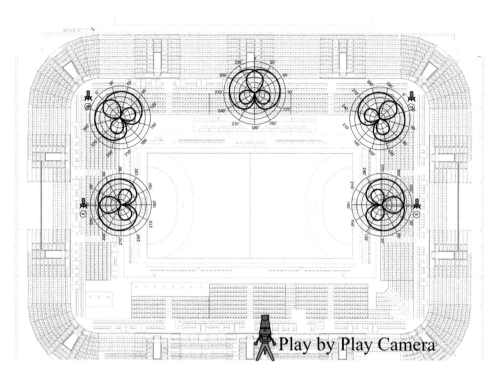

Figure 5.17 Variable spaced pairs: Placement to cover/capture the left, far and right side of the arena

design is more of a vertical coincident pair. Dr. Lee constructed a quadraphonic layer at the ear height augmented with another quadraphonic layer elevated at 30 to 45 degrees, cardioid or supercardioid microphones facing directly upwards. Remember, according to Dr. Lee the height aspect can be achieved with microphones mounted like his PCMA-3D design where each corner of the square has a pair of mono microphones in near coincidence or mid-side configuration.

The biggest difference in Dr. Lee's and my designs is that in sports I subscribe to a much wider separation of the microphones, both horizontally and vertically. The ear-level sounds are captured from microphones on stands in front of the audience and the height microphones are mounted above and pointing down and above the audience similar to a very large Hamasaki Cube. Because of the homogenous nature of the sound of sports crowds the sound appears to be correlated enough for sports and in-phase on a vector scope.

Synchronization

The necessity to have synchronization between the audio and picture is not an open discussion but I continue to see lip-sync issues almost daily. However with sports there is a big fudge factor when capturing ambiance and atmosphere. Additionally, electronic graphics and images commonly inserted over live video may require adjustments for synchronization.

Sound Capture: Additional Topics

Microphone accessories can be overlooked, underappreciated and perhaps even misunderstood.

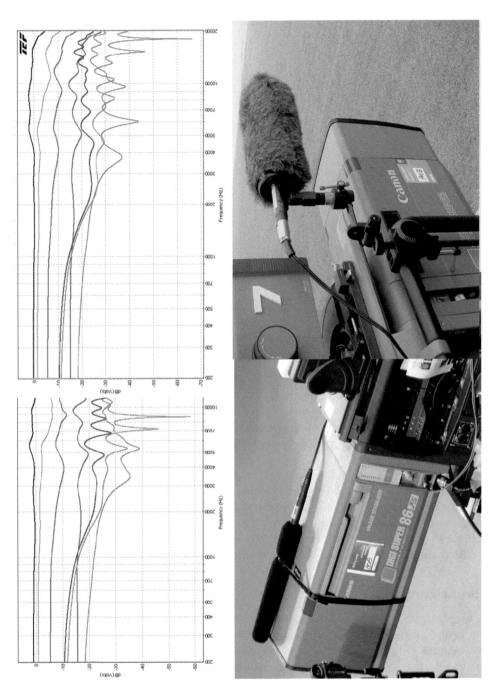

Figure 5.18 Graph of microphone placed directly on top of flat surface vs. free space measurement

Mounting and securing microphones in certain soundspace often requires stands and specialty mounts. For decades I have provided hundreds of stands for the audio technicians at the Olympics and was often surprised at their incorrect use or lack of use. It has been common for audio assistants to place a shotgun microphone directly on top of a camera lens. Well this kind of works because the top of the camera lens becomes a boundary surface but I was certain there was a significant compromise in performance.

I asked Dr. Helmut Wittek to measure the effect on a shotgun microphone by lying it directly on top of a camera lens as opposed to mounting the microphone a couple of inches above the camera lens on a stand. The difference in polar patterns is astounding.

Mounting microphones should be common sense for most audio practitioners. For example, football generally uses a pair of shotgun microphones on stands behind the goals. Short stands put the shotgun microphones too close to the ground, which is an absorptive surface and does slightly attenuate some capture – a mid-size stand, 1.5 meter, would improve the capture. When placing a microphone always be aware of the direction the sound disperses from PA cabinets and try to place microphones as off-axis to the PA as possible.

Colors

The directional boundary microphone is available in a complete and portable low-profile package in black and white coloring with matching cables making it useful in sports application because of its size, profile and color which usually blends into the background. Audio Technica even painted a couple of microphones for the Olympics.

Wind and Rain Protection

The best microphone is useless without proper protection against the weather. Wind and rain are the nemesis of a microphone and can render a microphone momentarily useless and rain will eventually make a microphone and most electrical devices and connections inoperable. Rycote has dominated the "wind shield" business with continued innovation and customer service under the leadership of Simon Davies, even matching the color of some windshields to Wimbledon green.

Notes

1 Lee, Hyunkook, "Multichannel 3D Microphone Arrays: A Review", *Audio Engineering Society*, vol. 67, no. 1–2. See also: Lee, Hyunkook, "Capturing 360° Audio Using an Equal Segment Microphone Array (ESMA)," *Audio Engineering Society* vol. 69.

2 Wittek, Helmut Microphones ""ORTF-3D": A microphone setup for 3D Audio and VR ambience recording," www.hauptmikrofon.de/3d/ortf-3d

3 Gerzon, Michael A., "Periphony: Height sound Reproduction", *Journal of the Audio Engineering Society* 1973.

    See also: Gerzon, Michael A., "Ambisonics" www.michaelgerzonphotos.org.uk/ambisonics.html; Gerzon, Michael, "What's Wrong with Quadraphonics?", *Studio Sound*, vol. 16, no. 5, pp. 50, 51, 56 (May 1974).

4 Jordan, Larry, "Brian Glasscock: Ambisonic Mics for 360° VR Sound", *Digital Production Buzz* www.digitalproductionbuzz.com/interview/brian-glasscock-ambisonic-mics-for-360-vr-sound/#.YYwEd07P1LI

5 EigenUnits® VST Plugins for macOS and Windows, Version 2 Rev. B October 2019 https://mhacoustics.com/sites/default/files/EigenUnits%20User%20Manual_0.pdf

6 Metering

The Art and Science of Seeing Sound

Mike Kahsnitz, CTO RTW, co-author

Arguably, controlling audio signals is the most serious responsibility for the engineers, audio operators and technicians working on a live production or in the studio and metering and monitoring the signals is essential to success. Monitoring and listening to the audio program though an appropriate speaker or headphones is crucial to achieving quality sound, however in addition to analytical listening, measuring these audio signals and providing adequate and informative visual references is fundamental. Metering and monitoring devices along with the supporting software are valuable tools for quick visual reference and trouble-shooting, but always remember these programs and boxes can let you look at a visual representation of the audio, but can never guarantee the fidelity and quality of audio content – only someone's ear can.

To maintain consistency in the audio from production to distribution, quantitative and qualitative evaluation and measurements of the sound should be maintained throughout a production and through the broadcast signal chain. Qualitative measurements like spatial imaging and phase are significant considerations for all audio productions, but conforming to the broadcaster's deliverables specifications for loudness and true peak across all mix formats is critical and mandated by law.

There is no single set of tools for immersive audio management. Audio management tools for metering, monitoring and mastering are common with all professional audio production, however the tools used by a live production mixer are certainly different than the needs and tools used by a post-production sound mixer.

How Did We Get Here?

A visual reference is a vital tool in a chaotic OB Van because an experienced sound engineer learns how to visually associate an image or graphical reference with the sound they hear. The visual reference may alert the engineer that there is a problem and that he needs to pay attention to something specific. The meter lessens difficulties that stem from poor listening conditions, stress and aural fatigue. Remember the sound engineer must listen to the director and producer, as well as create multiple mixes and in the case of sports and live entertainment in real time.

To understand immersive sound metering is to understand how audio has evolved to this point in an ever changing world of sound. As sound practitioners we use our hearing as an evaluation tool because our ears are a quick and accurate indicator of changes in loudness. Dramatic changes in the sound will instinctively trigger a reaction from a respectable live sound mixer to adjust an audio level somewhere.

Digital audio brought greater dynamic range and a greater level of difficulty for matching and maintaining program audio between different pieces of equipment. Basic volume/level metering is common to anyone who ever used an audio recorder, mixer or virtually any

DOI: 10.4324/9781003052876-6

Figure 6.1 Dolby's proprietary dialogue intelligence algorithm was built into the LM100 Broadcast Loudness Meter which uses ITU-R BS.1770-1 as its core measurement algorithm but users can select between weighted and unweighted peaks and more

sound software graphic user interfaces; they all have some sort of visual volume reference for the operator. However, many of these meters do not meet the professional standards for reliable measurements, particularly for digital and high-definition audio workflows needed for surround and immersive sound.

With digital and surround sound audio it quickly became apparent that there were significant ideological and production discrepancies between content producers and advertisers who were pushing the dynamic envelope with sonic enhancements. Much broadcast content is dialog driven so Dolby Labs developed a dialogue intelligence algorithm specifically designed to measure the perceived loudness of dialogue. Offensively loud advertising is usually dialog driven as well and now the broadcasters had a tool to put some limits on producers.

This basic screen and measurement documented the fact that consumers already knew – audio levels were all over place. HD and immersive audio brings a greater need for visualizing, not just of electrical parameters such as volume levels and phase relationships but also metadata and dynamic characteristics of the audio signal, and practitioners required a means of reliable measurement.

How Does a Meter Work?

The sound we hear and produce is captured and converted to an electric digital fingerprint that can be amplified, manipulated and distributed. As sound move between gain stages, processors and plug-ins, proper volume levels/volume management is a necessity to ensure good gain structure between the noise floor and to minimize excessive distortion in the electronics. Note: distortion can occur at almost any point in the audio path from poor gain structure.

Sound management for analog and digital signals begins with a method to measure the amplitude of the sound signals. Volume is a scientific measurement of the power of a sound and is measured using the decibel system. Many analog systems use the volume unit (VU) to measure and display the dB reference as a scale and not as a voltage reference. The VU meter often has a very limited, slow, sometimes inaccurate scale and only reads averages and not peak values.

When an audio signal is too strong for the circuit to pass, the peaks of the waves are flattened off, resembling a square instead of rounded like normal. This adds overtones related to the frequency of the original sound. Tube circuits produce a soft clipping when the signal reaches overload with a gradual breakup as the volume increases. Transistors produce a hard clipping and go from clean to distorted sharply. As a sine wave distorts it sounds more like a square wave. Distortion seems to be less objectionable with analog audio but digital has absolute peaks and once the signal reaches that threshold the resulting audio is unusable because the system cannot recreate that part of the signal.

When measuring audio signals there are two key parameters to consider – true peak and loudness. True peak enables the mixer to react to excessive volume or clipping while loudness helps the mixer manage the variations in the perceived loudness between different types of content and commercials. Conventional PPM (Peak Program Meter) are instruments that evaluate samples of the digital audio using their native sample rate. Unlike conventional analog meters which measure from the noise floor up, the digital meter measures from the clipping point down.

Adjusting and maintaining consistent volume levels using peak and true PPM type devices is manageable, but note that PPM meters do not provide effective loudness monitoring. Peak program meters give you a sense of dynamic range by displaying the maximum amplitude or volume levels of a signals waveform. However, peak program meters include a short integration period of between 5 and 10ms which deliberately ignore short transients.

Digital audio uses an absolute full-scale reference where zero corresponds to the maximum level. 0 dBFS (decibels full scale) is absolute peak and above this level is digital distortion. With such an absolute point of distortion, precise methods of measuring is critical with recommendations coming from the ITU BS.1770 for higher sampling rates to give a more accurate representation of the amplitude of the actual waveform. Higher sampling result in "true peak" level measurements. The Nyquist-Shannon Sampling Theory states that the higher sampling rate specified for true peak metering is necessary for an accurate measurement of loudness.

Measuring Real-Time Events

Peak meters and true peak meters are generally bar-graph type instruments which usually are linear and logarithmic and generally are a reliable indication of a wide dynamic range. Bargraph meters are an incremental horizontal or vertical display of Light Emitting Diodes (LED) which can be easily programmed with color schemes to aid the operator. IEC 60268-10 requires a minimum of 100 segments and a resolution better than 0.5dB at high levels easily implemented in computer software. LEDs can be programmed to emulate various meter ballistics and a range of meter characteristics. Color patterns can be used to facilitate easy viewing and interpretation. For example, in order to visualize the signal at the -9 dBFS level a color or brightness change should result.

Metering is the tool to assist the audio practitioner in maintaining proper gain structure and professional meters can be programmed for all practical implementations, particularly applications with near full-scale sound levels. There is a reason to include headroom in your audio measurement for live entertainment and sports broadcasting because beyond a certain full level digital saturation will occur which is problematic with equipment that has an extensive dynamic range. Remember, even levels that are low can be amplified without adding excessive noise while digital distortion is irrecoverable.

Information obtained from metering facilitates adjusting and maintaining clean undistorted signals either automatically or manually by a hands-on mixer.

Distortion can be introduced anywhere in the signal chain but meter and monitoring of the audio signals needs to begin at the source – in the control room or OB while loudness management and monitoring can be centralized down stream.

Loudness

The wide variety of material that is produced for broadcasting and streaming makes accurate loudness measurements a challenge. News, music, sports, talk and advertising all require different considerations to make accurate measurements. In the most fundamental sound pressure meter there has always been a system for weighting or filtering the sound for an accurate measurement. A Scale – B Scale – C Scale.

Dialog and Music

Additionally for content producers and broadcasters, loudness became a difficult problem as a result of the transition from analog audio to digital and an extended dynamic range of digital TV over analog. Loudness and volume are often confused but loudness is one of the most relevant audio measurements in broadcasting, simply because it is regulated by law. The volume battle between broadcasters, advertisers and content producers resulted in the CALM act – US legislation to ensure that TV commercials don't overwhelm program levels. The CALM Act only applies to terrestrial and other over-the-air distributors but not to streaming radio and Over the Top (OTT) video distributors.

Most TV devices today can easily switch between conventional TV that might be delivered by cable, satellite or streaming and of course streaming program and media servers. For this kind of distribution you want to be sure that the loudness is at least within a certain range from one program to another. For mobile devices used with low dynamic, poor speakers or used in a loud environmental situation you want to have a louder signal with lower dynamic.

Another standards body, the ATSC-85, makes a distinction between dynamic range and loudness and states that VU and PPM meters are not acceptable for measuring loudness.

Next Gen Audio Metering

It was necessary to develop an accurate measurement scale for objective loudness comparison of signals. Unlike electrical measurements, loudness is the way we subjectively perceive amplitude/volume and is based on factors such as frequency and duration and not just the amplitude of a sound. Consider that a change in amplitude is not necessarily perceived as a proportional change in loudness because our perception of loudness is triggered by frequency and timbre. The noticeable perception of change in amplitude varies by frequency and is well known as the famous Fletcher-Munson curves of equal loudness.

You were probably first introduced to this concept by the loudness button on your old stereo amplifier. This was a bass boost or low frequency boost for listening at lower volume because many natural sounds including music contains a lot of low frequency information which the listener often would enhance.

Loudness, as well as peak levels, needs to be measured at various points in the production and distribution path from the mixing room in the OB Van, soundstage, fiber and satellite paths, through master control and finally at various distribution conduits to the consumer. The time window for measurement is fundamentally based upon production. A loudness meter for live broadcast needs is entirely different from long-term loudness analysis of a broadcast station. Live events need to measure with a relatively short integration time focusing on the loudness of the anchor element – generally the dialog.

Surround sound complicates the problem of loudness because of the wide and varied use of the surround sound channels. Advertisers use the surround channels differently than sports, drama and entertainment programing. With advertising it may be common for sound elements to pop out from different channels.

With live productions, a functional and useful metering system needs to provide current loudness measurements in real time. This information guides the sound mixer in the management of the dynamic range of the audio content and facilitates mixing to a target loudness level.

After listening tests and evaluation the International Telecommunications Union (ITU) defined an algorithm to measure the loudness of 5.1 surround sound content. The ITU BS 1770 measurement technique uses the sum of five channels of the surround mix, not including the LFE. All five channels pass through the same filtering and a gain weighted summing process with their individual results scaled and summed together and give a loudness number.

The ITU loudness metering was initially scaled in dBLU – Loudness Units with a scale range from –21 to +9. The broadcast industry has generally settled on using a similar measurement formula with a different scale. Loudness is measured in negative units starting at –31 LKFS and counts down to zero. For example, the closer you get to 0 the louder the content and a measurement of –9 LKFS is louder than a measurement of –31 LKFS.

Live event measurements need to be made with a relatively short integration time with the loudness of the anchor element factored as the event progresses. With long-term measurements, special care should be taken to detect modulation breaks so that they don't alter the results unintentionally. This is provided by two gating functions in the algorithm. There's one absolute gate, the so-called Silence Gate, that kills all values below –70LKFS and a second gate called relative gate that, as the wording says, reduces the measurement data from values that are out of the range of –10 LU compared to the integrated value at the actual time.

Dialog-based content is dominant in the mid-frequency range while music has greater frequency content. The K refers to applying an equal loudness contour (Fletcher Munson Curve) to each channel to ensure the measurement corresponds to the subjective perceptions of loudness like the brain does.

The A1 (sound mixer) must monitor and digest a tremendous amount of information from production commands, a director, six speakers, meters and dials, plus deliver a quality show mix. A metering system for "live production" must provide multiple information windows for true peak information, loudness and more.

Since loudness compliance for a live production is dynamic and often a manual process, live mixing desk manufacturers from Lawo and Calrec include many audio measuring options and capabilities directly into the console for easy viewing and screening by the operator.

Compliance Metering

Long-term measurements of loudness values are useful for broadcast control, QC and subsequent program analysis. Such a tool may be used for examining the loudness history of the program over several hours or days. The measurements can also be presented as histograms that sort the magnitude of the measurements into 1 dB or 1 LKFS-wide slices and show the number of measurements that fit into each of these slices. The histogram thus provides an effective portrayal of the consistency of the loudness in compliance of the CALM Act to ensure commercials do not overwhelm program levels. This again allows for visualizing loudness uptrends or downtrends that the engineer can then compensate manually. During this process, an ongoing (dynamic) time window is critical to ensure that averaging always occurs over an identical time range.[1]

Another useful programmable function is a running measurement of the production loudness from its start to any point to provide an indication of the average loudness of the production to the current point of interest.

Let's consider techniques for measuring short term. For example, an evening talk show has an opening monologue with lots of audience reaction then the band plays and takes the viewer to commercial break. In addition to evaluating the time constant and frequency there must be consideration for the gaps in content. Gating and threshold features serve to pause the short-term loudness measurement when the signal drops below a certain threshold. Threshold setting and user adjustable gating are also useful for dialog-based programming which may contain long pauses and skew the loudness measurements. The key to appropriate loudness measurements is setting the adjustments so that only the relevant parts of the program are measured. There is no gating for medium- and short-term measurements, only for integrated program loudness in this standard. Also the gating parameters are not adjustable for a standardized measurement. Our devices allow the gating to be modified only in a user mode. That was more or less created for very experienced engineers for testing purposes, but is not allowed for standardized measurements.

The RTW meter, popular with most broadcasters, is a comprehensive display with loudness bargraphs and true PPMs for each mono, stereo, surround and immersive signal path along with a surround sound and immersive sound analyzer plus spectrum analyzer.

Figure 6.2 The RTW meter is a comprehensive display with loudness bargraphs and True PPMs for each mono, stereo, surround and immersive signal path

Many of us are familiar with the RTW "House" display, which visualizes phase and loudness relations between the channels. RTW built on its familiar surround sound analyzer ("house" design) now, two stories. When the house display shows a square, the four channels L, R, LS, and RS share the same sonic pressure level. If the sides lines (vectors) are straight and do not bend then the individual channels do not correlate. Boundary lines bent outward indicate a positive correlation of two channels while lines bent inward show a negative correlation; that way an inverted phase can be easily identified and corrected. One of the sides being shorter indicate a missing or low level.

The yellow line (vector) of the center channel is interlinked with the L and R front channels by separate yellow lines. This allows for quickly realizing the relationship between the phantom source formed by the L and R channel and the center channel.

The RTW loudness display offers various scales for loudness calibrated, i.e. in LU from – 21 to +9 or LKFS from –31 to 0 or others with selectable reference values, integration times, and threshold setting. Advance features for dynamic time windows and user-adjustable gating ensure only the relevant parts of the program are used for calculating loudness values.

Phase correlation between the channels are useful measurements when ensuring downmix compatibility between surround, stereo and mono, as well as identifying any phase anomalies between channels and assist in optimal microphone placement. Downmixing out of phase signals will cause drastic sonic changes due to phase cancelation.

The correlation display will operate over a wide range of levels and does not affect the accuracy of the display values. The correlation scale ranges from –1 to +1 with a 0 value in the center between them. Additionally, the correlation meter can give an indication of the width of the stereo signal. Entirely identical signals have a correlation of +1, unrelated signals have a relationship of 0. When two channels of a stereo signal are identical but out of phase due to a 180-degree polarity, reversal will show a value of –1. A value between 0 and –1 may result from phase modulated components by effects processing.

Advanced Immersive Sound Metering

Advanced metering for surround and immersive sound not only requires a need to measure electrical parameters but also to look at phase relationships as well as the spectral and spatial characteristics of the audio signal. With the incrementally larger numbers of channels used with dimensional sound comes the increased risk of problems from channel failure and drop-out as well as level and phase issues between channels which are fundamental for high-quality dimensional sound and need to be constantly monitored.

RTW combined the features of two surround sound analyzers into one system for fast and intuitive visualization, something that was already familiar to audio mixes and also easy to understand. RTW wanted to stay close to their unique surround sound analyzer, which a lot of audio engineers were already familiar with and retained some of the features.

The RTW immersive sound analyzer is principled on its established "house" visual reference, but the immersive sound analyzer expands the principles and creates the immersive group from visually concentric beds – the main bed(s) and the top bed(s). In most immersive groups you add at least one 5.1 (or 7.1) main bed and one 4.0 top bed or more. Note: the immersive group will overwrite any settings in the sub-audio groups for the two beds to guarantee correct operation as well as set the overall weighting factors.

The main bed includes 5.1 or the 7.1 part of the format and the upper bed contains the 2.0 or the 4.0 part of the immersive format. This approach monitors the relationship between the two layers and easily lets the audio practitioner keep an eye on the overall surround image balance on both main and upper beds. Immersive sound metering requires at least one audio group for the main bed and another audio group for the upper bed, but can accommodate more.

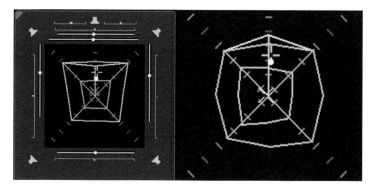

Figure 6.3 Main and upper bed combined. Here you see the combination of the main bed and upper bed, including the IDI dot. ISA upper and lower bed

With new immersive sound formats such as Dolby Atmos, the audio world is having an increased focus on audio formats with a large number of audio channels. RTW told me that the challenge is to measure and display all of these channels in a way that is usable in the many immersive audio applications. The immersive sound analyzer is RTW's and MK's vast sound experience designed into an instrument that can get you relevant technical and subjective parameters of formats such as 5.1.4 or even 7.1.4 in a fast and intuitive way.

The examples in this publication will be based on a standard 5.1.4 Dolby Atmos setup.

The Components of the Immersive Sound Analyzer (ISA)

First, let's have a look at each individual component (group) of the immersive sound analyzer. There are groups/components for each of the main and upper layers and there are metering functions that shows the relationship between the layers. The parameters of the upper bed are displayed with orange-colored lines which optically move closer to the viewer. The orange area shows the volume and balance of the upper bed. The dominance indicator (DMI) crosshairs is tilted 45 degrees. This is to maintain a clear distinction between the main and upper bed DMIs.

The immersive sound analyzer can detect the total volume and balance of the main bed by the total volume indicator – the cyan-colored bargraphs coming from the center and its surrounding lines. The (blue) crosshairs is the so-called dominance indicator (DMI) that shows the position of the subjectively perceived acoustic focal point. In this layer you can measure the total volume, loudness and focal point easily.

The relationship between the main and upper beds is indicated by the size of a white dot, as you can see here. The white dot sits between the main and upper bed dominance indicators. This perceptive feature is an immersive indicator of how the main and upper beds relate to each other. The dot is called the immersive dominance indicator (IDI) and the position and size of it indicates the loudness relation between the main and the upper bed. A very small dot indicates that there is audio only in the main bed, and a large dot indicates there is audio only in the upper bed.

The perceived position and its height, the ambient signal of the two beds and the spreading and phase correlation of a selectable bed can be displayed simultaneously in one easy-to-read view. The total volume indicator (TVI) displays the total volume and balance of the audio material in a single view.

Once familiar with the display, you get a fast and intuitive overview of a 5.1.4 signal. The perceived position and its height, the ambient signal (TVI) of the two beds and the spreading and phase correlation of a selectable bed can be displayed simultaneously in one easy-to-read view. The total volume indicator displays the total volume and balance of the audio material in a single view.

The immersive sound analyzer shows the surround images for both the main and upper bed in one easy-to-read window. The immersive sound analyzer displays total volume indicators for both beds simultaneously, and the position and height of the dominant signal is shown by a white dot. Additionally, the immersive sound analyzer can help the audio practitioner identify phase problems to make sure your audio will down mix correctly.

You would also need to keep an eye on the correlation and distribution between the audio channels in both the 5.1 layer and the 4.0 layer. By using the immersive sound analyzer, you can quickly spot audio material that is not balanced correctly plus you can keep an eye on the subjectively perceived acoustic focal point.

Beside PPM/True-peak and loudness, the channels outside the 5.1.4 main and upper bed (objects) should be monitored with regards to image and phase problems. The immersive sound analyzer will indicate phase problems and show you the stereo width between each of the speakers within their bed.

It is generally recommended that the main and upper beds, as well as the additional stereo channels plus the objects, are all monitored on momentary and short-term loudness as those measures give you a clear picture on the perceived levels of each group.

The phase correlation bars display the phase relation between the different sources of the selected bed as well as phase correlation between adjacent channels. Like the phase correlation bars, the phantom source indicators (PSI) show relations between each of the channels in a surround mix of the selected bed. If you check the settings of the two bed audio groups, you will find that some settings are now greyed out and locked. Additionally, there is a LFE phase warning in case of negative correlation between any channel and LFE is detected.

Now let us take a closer look at what it makes sense to monitor if you work with a typical immersive audio stream in broadcast. This example will show a 16-channel screen layout according to 5.1.4, one stereo pair and four objects, but it can of course also be done with two stereo pairs and two objects, or even with 7.1.4 based audio layouts.

Below you see a representative immersive sound metering configuration that covers virtually all technical requirements for a 5.1.4 immersive sound program plus additional visual monitoring for a stereo mix and four mono objects – totaling 16 input channels to the meter.

Across the top of the screen includes true peak levels for each audio group – the bottom 5.1 (left), the top 4.0 (middle) and stereo metering with loudness metering and adjacent peak meters along with a vector scope (right). Below the peak and loudness metering is an immersive sound analyzer with loudness metering next to the "house" graphic. In the lower middle section of the screen is the alphanumeric readout for momentary, short term and integrated LUFS. The right section is reserved for information about the audio objects (elements) plus another alphanumeric display and function buttons in the bottom right-hand corner. True-peak levels for the bottom group. Note: this meter is directly above the immersive sound analyzer and easily associated with the analyzer.

In most production scenarios, engineers need to keep an eye on the PPM level per channel, including true-peak warnings. Naturally, it is important to monitor if there is in fact a signal present on the channel, but also, and this is equally important, to monitor whether audio overloads occur. Overloads can create a large number of problems downstream, so it is important to keep an eye on this aspect constantly. The white line on top of each bar is a TP max marker that makes it easy to identify TP overloads on specific channels.

Worldwide regulations (EBU R128, CALM Act and more) require broadcasters to guarantee true-peak levels within specific limits for each audio channel. There are a number of international standards for measuring stereo and surround audio. ITU 1770 in its latest version 4 (ITU-R BS. 1770-4) covers the most popular configurations and weighting factors for measuring audio

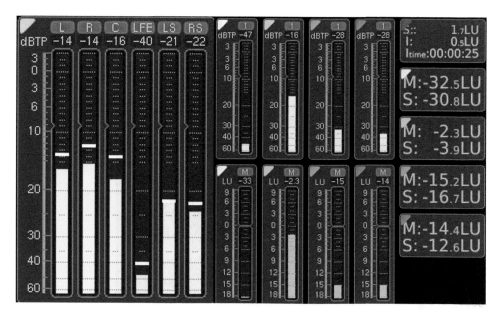

Figure 6.4 PPM view

layout formats up to 22.2. The RTW PPM meter fully complies with all international standards. All audio groups should be monitored with regards to various loudness measures.

There are a number of ways to display momentary and short-term loudness, including a PPM-style meter as the one you see in graphics.

LUFS – Loudness units relative to full scale.

LKFS – Loudness units K weighted relative to full scale.

LU – Loudness units relative to the reference of the chosen standard (ITU 1770 i.e. 24 LKFS or EBU R128 -23 LUFS)

Momentary Loudness is measured in a time window of 400 ms.

Short Term Loudness is measured within a time window of 3 seconds.

Integrated – perceived loudness over the entire program from start to finish.

LRA – Loudness Range can be associated with the dynamic range – it is the difference between average soft and average loud parts of an audio segment.

Tpmax – Max True Peak

Mmax and TP-hold

Loudness across the 5.1 main and the 4.0 upper bed.

The Vector Scope

In addition to the immersive sound analyzer, the vector scope is one of the most fundamental tools for the serious audio engineer. The audio vector scope is used to monitor the balance of stereo mixes, check the stereo width and perceived position between the channels plus display the changing phase relationship between a selectable channel pair in real time. The behavior of the constantly moving Lissajous figure and its spreading may give information on the width of stereo base or if there are comb filter effects or phase shifts or rotations in the signal. To keep the display readable, and to be able to work with a wide range of levels, the vectorscope has a built-in automatic gain control (AGC) that will compensate for changes in levels.

This picture appears well balanced and typical of a stereo mix.

Here is an example of how the lissajous dislay looks when the signal is close to mono.

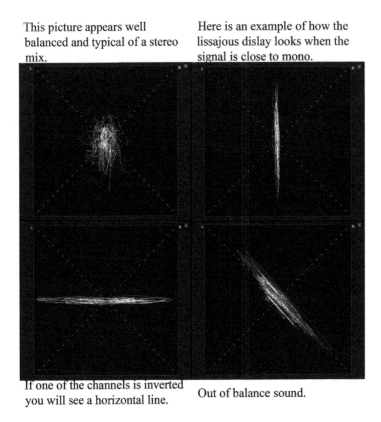

If one of the channels is inverted you will see a horizontal line.

Out of balance sound.

Figure 6.5 Vector scope patterns

Dual-mono signals with exactly the same level in each channel appear as a vertical line with level-dependent deflection. Generally, vertical patterns mean that left and right channels are similar, so the more vertical the pattern is, the more you approach mono.

Sources that would normally be very close to mono are those traditionally recorded in mono and panned center: dialog, lead vocals, lead instruments, bass, kick drum etc. Adding effects like stereo reverb will of course spread the source in the stereo image. Watch out for horizontal patterns. Generally, horizontal patterns mean that the two channels are very different, which could result in mono compatibility problems, and therefore you should pay attention if you see these patterns. In many cases, switched polarity occurs due to wrong cabling or if the INV button on a microphone or mixer is unintentionally switched on.

Tilted stereo image signals that are present only in the left channel are displayed as a line rotated by 45° to the left, while signals present only in the right channel are rotated by 45° to the right. This is how it looks if the stereo image is out of balance and the signal is mostly in the left channel.

For surround channel configurations (4.0 and above), the vectorscope can be switched to a four-channel mode. In this case, the L/R channel pair is displayed only in the two upper quadrants while the two lower quadrants show the phase relationship between the LS and RS channels of the surround signal.

Mono objects: as mentioned earlier an audio group can be comprised of as little as one – a mono channel – to as many as eight channels as used in the bottom group of a 7.1.4

configuration. An object is audio that can include spatial and spectral information requiring that each object be monitored with regards to true-peak PPM.

Not knowing what the final mix at the end user's place will be, for most productions you would want to ensure consistency between the objects to avoid loudness jumps and other artifacts, so the loudness measures are key on this setup.

Finally, the spectrum analyzer can help make decisions as to whether there needs to be alterations or enhancements of any frequency or band of frequencies. The spectrum analyzer can be programmed to provide a sound practitioner the spectral balance of a program to assist in adjusting EQ as well as trouble-shooting and compensating for sonic problems such as resonant frequencies and even irritating frequencies from a HVAC.

Designing the Immersive Display: Touch Monitor Screen-Layout Editor

The RTW touch monitor system is highly flexible with regards to designing the screen layout to create your own visual preferences and comfort. Since the final display is a composite of the various audio groups and functions that have been discussed before, then you need to make sure that you have created the necessary audio groups and set them up correctly before designing the actual arrangement and display of the audio groups and function displays. Alphanumeric references are included for precise compliance measurements.

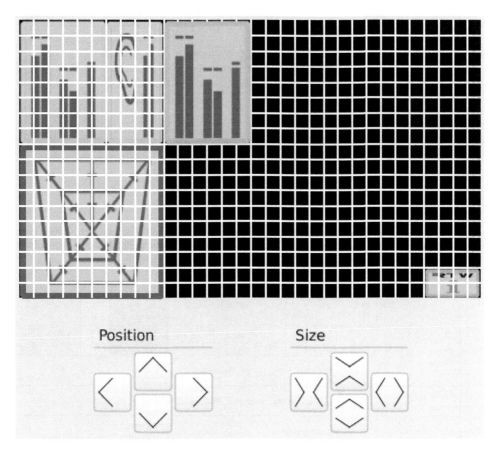

Figure 6.6 The TouchMonitor screen-layout editor lets the audio practitioner arrange and organize the screen information on a grid with position and size controls

The stereo section shows true-peak PPM as well as momentary, short-term and integrated loudness as graphical bars. Further, I would recommend using the vectorscope to let you keep an eye on the stereo image as well as the phase. Like the stereo pair, I have set up only a momentary loudness graphical bar, and I have added momentary and short-term loudness as numerical values for the mono object channels. Additionally, each audio group containing the objects is marked with different colors for a better overview.

A set of START, STOP and RESET buttons are always convenient to have for starting and stopping measurements and to reset loudness calculations. The buttons shown are part of the global keyboard and linked through all audio groups to synchronize start/stop of the integrated measurement as well as to reset all measurement stored in memory.

This is a basic display of relevant technical and subjective parameters of immersive signals in a fast and intuitive way. Useful for immersive live broadcast production, as well as monitoring the broadcast transmission path.

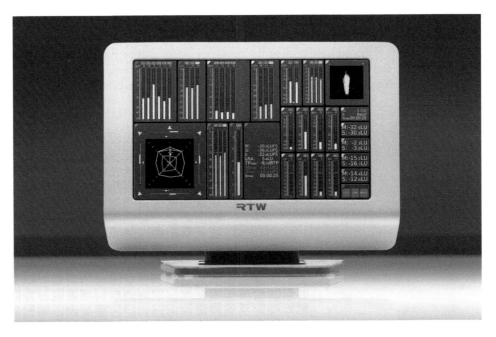

Figure 6.7 Full view RTW screen. Here you see the final screen layout displaying the entire immersive audio stream, combining all of the above. A screen layout perfectly and personally tailored for immersive audio metering

Main Bed 5.1						Upper Bed 4				Stereo 1 2		Obj 1 1	Obj 2 1	Obj 3 1	Obj 4 1
1 L	2 R	3 C	4 LFE	5 LS	6 RS	7 TL	8 TR	9 LTS	10 RTS	11 L	12 R	13 M	14 M	15 M	16 M
PPM, True-peak															
Momentary Loudness, Shortterm Loudness															
Surround Image										Stereo Image					

Figure 6.8 RTW has presets for typical standard channel assignments

Live metering is a matter of preference and need. A lot of information in a crowded space may not be as useful as switching menus; besides button pushing is something the audio practitioner is good at.

Post-Production

The metering and monitoring tools used in live production and in post-production are completely different in their implementation and operation. Basically, the RTW meter operates in real time with no latency, while a plug-in and the DAW require some processing time for rendering – although this can be negligible. Audio post-production usually occurs on a digital audio workstation (DAW) where the audio practitioner works with files and not digital streams. In the post world the audio operator often uses sophisticated third-party software tools and plug-ins for accurate processing and referencing. A significant difference between the plug-in, DAW and the RTW meter is that the RTW meter is transparent to the audio signals while many of the plug-ins and the linear acoustics and Junger audio black boxes are designed for additional processing including processing the dynamic range of the audio.

There are a variety of loudness tools that can be applied to an edit session to meet a content provider service's requirements. Nugen supports native 7.1.2 and 5.1.4 audio processing and loudness parameters for advanced loudness control and dialog consistency. Dynamic Range Control (DRC) is nothing new and along with the equalizer, the compressor is one of the oldest audio processors. The DynApt is an optional extension for Nugen's LM-Correct which allows corrections with dynamic adjustments, not just linear adjustments. This means the user can correct multiple sets of parameters simultaneously and gives the extra option of narrowing the LRA (loudness range). DynApt has a unique algorithm which uses volume-riding and scene-change recognition to make transparent-sounding adjustments.

During rendering, the software conforms the mix to the required loudness specification and prepares the content for delivery over a wide array of audio formats from mono to various immersive formats supporting up to 7.1.2. Nugen's software can also down-process audio signals with its Halo Downmix feature that gives the audio mastering process new ranges for downmix coefficients, and a Netflix preset as well.

The Nugen visualizer is a metering and monitoring tool that can be used as a plugin during mastering and as a stand-alone program on another computer. The visualizer analyses qualitative characteristics of your audio that impact the clarity and value of your sound. Phase and frequency correlation impact the dynamics and spatial imaging of a mix and sometimes it is what you do not hear like low frequencies that pass through unheard that ruin a great mix.[2]

Beyond audio fidelity, proper evaluation of sound in the broadcast environment can be a difficult process because circuit continuity and signal routing can take precedence over sound fidelity in a chaotic broadcast situation. Sometimes the sound production process accumulates noise and distortion along the way, something a skilled sound mixer or tonemeister might hear but a transmission technician might not.

The Izotope Insight has incorporated the Netflix Broadcast specifications as well and supports Dolby ATMOS 7.1.2.[3] The Insight provides all the requisite loudness information but also a visual reference of frequency spectral information. Of interest to broadcasters is the intelligibility algorithm which monitors dialog in the context of different noise levels in a listener's environments. This is selectable between low, medium and high. The source meter displays the perceived loudness of the source. The thin white line above the dot is a history meter that displays the range over the course of 10 seconds. The difference meter displays the estimated difference between the perceived loudness levels of the output and the selected source. In addition to intelligibility the device can display the percentage of dialog detected in the source material over the entire duration of a gated loudness measurement. The measurement represents the ration of detected dialog content and total content in the source material. The calculation is made every 500 milliseconds. The Izotope Insight is a passive device that by itself does not process any audio but can send metering data and receive information through the inter plug-in communication (IPC) to other compatible Izotope plug-ins. The Izotope Insight is capable of real-time operation but requires a platform to operate on, such as Pro tools, Reaper, Nuendo and others.

Notes

1 Calm Act 2009, H. Rept. 111-374, 111th Congress (2009–2010) www.congress.gov/bill/111th-congress/house-bill/1084/all-info.
2 "Nugen Interviews with Freddy Vinehill-Cliffe, Solutions Specialist", https://nugenaudio.com/search/Freddy+Vinehill-Cliffe (accessed 15.12.2021).
3 "Netflix Sound Mix Specifications & Best Practices v1.2 – Netflix | Partnerhelp" https://partnerhelp.netflixstudios.com (accessed 15.12.2021).

7 Monitoring
The Art and Science of Hearing Sound

Not only must future ready entertainment consider the production side but also the consumption side of enhanced entrainment features because if the consumer can't experience immersive sound or interactivity then there will be no demand. Ceiling mounted speakers are generally out of the question for the home making reflected reproduction with up-firing and side-firing enclosures a probability. Producing immersive sound for general consumer consumption is difficult because there are so many different manufacturers producing different products with seemingly conflicting reproduction possibilities. For example, Dolby Atmos is everywhere, from the cinema where it was born to soundbars and up-firing speaker configurations.

How can all these different products deliver true dimensional sound? Sound reproduction from above the listener is the most significant change/addition to surround sound and was an easy implementation for the theatres that had sufficient wall height. Speakers along the walls, in front and in the ceiling produced a soundfield from every direction that engulfed the viewer. However, Dolby Atmos for home advocates bouncing the sound off the ceiling from speakers placed slightly above ear level, while DTS uses existing speakers and direct radiating speakers to reproduce height. DTS has an advantage because it works with standard surround sound setups.

Discrete immersive sound uses a combination of ear level speakers and either two, four or six height effect speakers. How can you create immersive sound with up-firing and side-firing speakers?

There are televisions, tablets and smartphones that support Dolby Atmos. I am trying to understand how these products can deliver an overhead dimension. DSP enhancements are common, rendering can adapt to the audio engine and ear devices so clearly there is an enhanced connection, but this aural experience is probably not the artistic intent of immersive sound designers.

The Sound of Sports and Entertainment

Rendering in the device is intended with Next Generation Audio but, at the time of writing, I have seen no real smart interface between an AV Device and speakers.

The AV device is usually the amplification system and includes decoders such as Dolby Atmos and variations of DTS. Consumer AV systems often include up-conversion of surround content to immersive using Dolby Surround and Neural-X. Up-conversion adds height information through DSP and the AV unit provides the channels and outputs.

With so many consumption formats and processing the audio producer and sound designer must have a reference basis as a foundation to build and create from. The sound practitioner must listen to their work through the capture, creation and mastering stages with different monitoring requirements and conditions.

DOI: 10.4324/9781003052876-7

Figure 7.1 Reflective reproduction is unpredictable because it is completely dependent of the home architecture

The dimensional dilemma is an issue for the audio practitioner. Monitoring, listening and mixing for direct height imaging versus reflected height brings concerns for the audio mixer, practitioner and sound designer. If a sound design cannot be accurately and reliably reproduced on virtually any system then the sound design is faulty or has specific reproduction requirements such as binaural ear devices.

The most reliable approach for the soundperson to monitor sound is with accurate speakers together with reasonable placement and using their own trained ears. Critical listening in small and acoustically challenged spaces can be tricky at best and almost impossible in an OB Van. The problems range from poor speaker placement, improper speaker calibration and alignment, poor acoustic treatment in OB Van, extreme ambient noise including excessive chatter on PLs. All this results in mixer fatigue and possible problems in the mix. Most OB shows must be mixed in difficult listening environments, but fortunately there are many skilled sound mixers that learned to adapt.

The adaption and evolution process began in remote production with the universal acceptance of 5.1.4 immersive sound for sports. This format dropped the middle speakers that are popular with cinema simply because they were too difficult to monitor, especially in the Studio and OB Van. Part of this adaptive process has been incorporating the use of visual monitors – not only for levels but also spatial orientation, correlation and phase. Keeping the live presentation manageable and scalable is adapting the expectations. Sports sound is not particularly as exciting as a cinematic production, but consider the production environment and consumption limitations of most live broadcasts, particularly live sports. Who is listening and how? Obviously, the best way to improve the sound of your mixes is to record them at home and listen later. As sound specialists learn and experiment with new sound formats the results will become more ambitious and scalable for a very wide range of consumers and users.

The final leg of the journey for better quality sound continues with educating the listening and screening habits of consumers, which already seems to have traction with the adoption of additional speakers outside the television frame and more television remote features. The television sets squeezed every centimeter of space out of the front panel and left no room for front-mounted speakers. Low fidelity rear-mounted speakers are standard but in defense of the manufacturers they assumed that the sound would be off-loaded to an ancillary system.

There is no doubt that soundbars leveraged the acceptance of surround sound and are crucial to the acceptance of immersive sound. The question that bothers me is that if our consumers are listening on soundbars, should the audio mixer be mixing on soundbars? No, it's not practical, but at a minimum someone should be QC'ing on soundbars.

Listening to Your Mixes

Listen to your mixes is the golden rule, but the question is how and over what? Particularly when it comes to immersive sound productions in confined spaces with acoustic anomalies and don't even ask, but over what sound format? In my mind the most significant consideration is over what speaker configuration and format you should listen to and this should be at least one standard that everyone agrees too. There are ITU International Television Union standards as well as recommendations from universities and research organizations like Fraunhofer for discrete speaker placement, but there is no standards for soundbars and their sonic reproduction.

Real vs. Virtual Speaker Configuration

All speakers, including height placement, should be equal distance from the listener, essentially in a hemispheric configuration as shown in Figure 7.2.

In many situations this is impractical and there are alternatives. There is an argument that the audio mixer should monitor in a standard setup as a basis for any subsequent dimensional mixes. I believe the rationale of this line of thought, but also believe that no matter what there will be compromises in the sonic quality and producer's intent in the sound between various formats. How can you bake in the dimensionality of an immersive mix into a stereo mix. Nine speakers vs. two speakers or headphones. You must listen to a mix over a variety of speakers and channel formats to get realistic feedback, particularly on spatialization. Sennheiser partnered with Fraunhofer to produce a single unit immersive soundbar that is being pitched as a reference monitor for home audio evaluation.

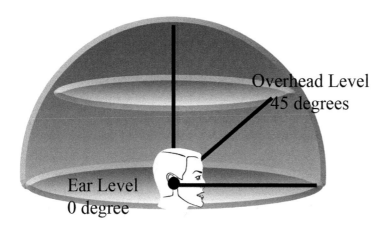

Figure 7.2 Speaker setup vertical layer – 45 degrees above the listener

I listen to a lot of recorded music from all periods of technology and different recording process and techniques. I am astonished at the wide range of sonic productions. It's the same thing with sports. Why does one production of a sport sound different than another production of the same sport. Certainly acoustics plays a factor, but surely inadequate and inaccurate monitoring in the OB Van plays a role.

As audio practitioners we are responsible for delivering an incrementally better experience while in the search for the complete illusion. We are adapting and learning in a liquid situation and must accept that the role of the sound designer is changing as well and might now include managing and programing the assets of the soundtrack for user control and manipulation.

Monitoring and Hearing

Individuals interpret what they hear differently because their ears and brains differ physically but also because of conditioning, training and filtering. The human ear can be trained to pick up subtle details in the soundfield, such as an imbalance between levels as well as distortion and phase shifts. However, everyone's ears and hearing is different which can impact the interpretation and course of action for an audio practitioner. For example, a person with high frequency hearing loss may miss distortion in the upper frequency ranges or may even compensate for the perceived lack of presence in a mix.

Because our ears are positioned on each side of the head and are focused forward, our hearing is particularly sensitive to sound that arrives directly in front of us. We also hear the differences between the lateral reflections from the sides. The brain measures the timing and amplitude differences in the sound as it arrives at the left and right ear. The timing differences of the sound to the ears help determine the direction and location of the sound source while the amplitude differences of the side reflections give a sense of the size of an acoustic space.

Any sound that is generated off-center has to travel further around one side of your head to reach your ear than it has to travel to the other side of your head. The head related transfer function (HRTF) describes the acoustic interaction between the sound wave and the listener's head, torso, outer ear and ear canal and the brain's resulting interpretation of direction.

From a sound practitioner's standpoint, human ears are most accurate at localizing sounds. The ears use difference in amplitude and timing to determine the location of a sound. The smaller the differences mean the sound is coming from the center front – the nose is roughly the mid-line. Sounds with greater differences are easily located but there can be ambiguities along the cone of confusion.

Sound emanating from the front and back, top and bottom along the cone of confusion will have the same loudness and timing differences and be difficult to precisely localize beyond if the sound is clearly to the left or right. While we are confused by sounds along the cone, humans are still able, to some extent, to determine the location of the sound along the cone because of the shape of their ears and head.

How Is the Consumer Listening to Your Mixes?

The migration to surround was a challenge for monitoring/listening to the sound for the audio practitioners as well as the consumer electronics industry. The professional audio producers tried to adhere to the recommended speaker setup, which was led by the movie and other post-produced audio producers who had the proper mix facilities while the audio practitioners in the OB and audio vans came as close as possible.

The change factor for surround sound was the soundbar because it became quickly apparent that the home and family setting was not going to accept five speakers in the living room

Figure 7.3 A typical 2.1 soundbar

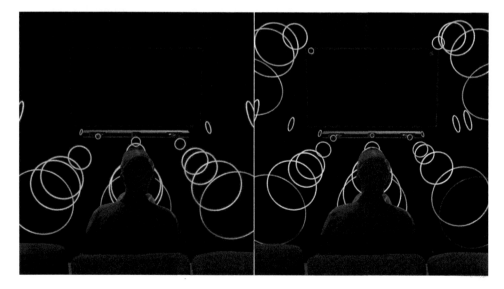

Figure 7.4 Samsung retails a soundbar that is wired to the television screen

much less the bedroom or the bathroom with their accompanying stands and wires. Without the demand for surround the format was as doomed as quad sound. The soundbar quickly proliferated with a range of offerings. The basic soundbar began as a single "loaf" housing, basically LCR – left, center, right, forward projecting speaker cluster. This was simple to set up and offered noticeable improvement.

Large soundfield reproduction over relatively few transducers is growing in sophistication and products like the Dolby Atmos Virtualizer and DTS Virtual-X are primarily for soundbars that use advance algorithms and psychoacoustic principles to virtualize the height effect.

This soundbar has front, side and up-firing speakers as well as the television speakers and does give a different level of immersion because the television speakers provide discrete height reproduction and not just reflected sound. Each soundbar device is capable of different levels of immersion with the Dolby Atmos and DTS:X codex/software resident in the boxes. The user tells the delivery renderer the channel configuration: 3.1.2, 7.1.2 or 9.1.4 and the delivery renderer optimizes the channels for the system.[1]

In an attempt to adhere to some standards for surround sound speaker configurations and placement, rear satellite speakers were added but the manufacturers encountered resistance even from sound enthusiasts about setup and wires. Remember powered speakers require electricity in addition to the audio signals. There was a clever solution that made the low frequency effect (LFE) wireless and then ran wires to the rear speakers.

Left, center and right speakers in the front soundbar configuration with real surround speakers reasonably reproduced the surround image, but soon came numerous offerings of surround capable soundbars in a single bar configuration. Single unit soundbar at their basic level are fundamentally front firing; however, there have been further refinements with up-firing and side-firing speakers that use sophisticated psychoacoustic DSP spatial algorithms and take advantage of room reflections.

If the soundbar is provided the audio and metadata for proper replication, the renderer will optimize the audio to the individual design and construction of the wide variety of speaker bars on the market now and into the future. The renderer is part of the decoder built into the soundbar.

Sennheiser and Fraunhofer partnered to produce a single soundbar unit that projects sound virtually in every direction – this is front-, up- and side-firing and is capable of recreating the illusion of spaciousness from material produced in immersive sound.

Although I subjectively like the quality of the soundbar, I believe it is too high fidelity to be representative of the consumer market which is why basic quality control should include low-end units. The soundbar market ranges in price and sophistication and precision reproduction which has complicated the mix process. Of significant consideration is that there is currently no meaningful evaluation for soundbars, but it seems sometimes the consumer could hear better than the audio mixer.

There are methods to improve spatial perception of soundbars by additional speakers in the rear and by augmenting surround speaker systems with up-firing boxes that usually sit on top of the existing speakers. Directionally firing speakers require reflective surfaces to function and discrete speakers require substantial installation; all will sound different in similar environments so the question is the consumer commitment.

How Does the Audio Practitioner Listen to Their Mix?

Many broadcast production spaces are an open floor plan where the director, producer, graphics and playback operators are grouped together. Generally, the audio mixer has a separate space; in older fixed installations, these spaces have survived the transition from stereo to surround but many I have seen have inadequate height for immersive. In this environment the only person monitoring the sound mix is the audio practitioners since the director, producer and everyone else in the control room is wearing headsets.

In the last 30 years, I have seen improvements in the OB Van audio space, but even with these improvements many spaces still fall short in adequate sound field reproduction. Generally, the audio space is too small and suffers from acoustic anomalies for proper monitoring. Room modes are resonances that occur when a room dimension is an exact multiple of the half wavelength of a sound being played back. A typical rectangular room has three main room modes corresponding to the dimensions of width, length and height. This means that there will be three frequencies at which you will expect to experience "hot spots." However, there are also a whole series of modes at frequencies above this corresponding to multiples of the three main mode frequencies. In practical terms, this means that some frequencies will sound much louder than others – even though the levels measured at the loudspeaker may be the same.

Many mix spaces lack proper acoustic treatment which causes interpretation of the spectral content of a mix to be problematic. Low frequencies in the mix are difficult to infer because there is no space for bass trapping while high frequency treatment has traditionally been something soft and fuzzy. There is a lot of use of acoustic panels like the Aurolux or panes built of Owens Corning Soundsoak. These panels are effective in their treatment of a wide range of frequencies but are not particularly effective in lower frequencies where problems like low frequency rumble can move a lot of air and not produce much sound.

Excessive low frequencies easily build up with lots of open microphones making a spectrum analyzer a valuable tool to insure against bass buildup. I find that many broadcast productions have a lot of bass frequencies which I believe is because of inadequate monitoring facilities. The sound mixer just cannot hear the low frequencies and often does nothing about it.

The OB environment continues to be a significant challenge because of inadequate isolation and noise. Isolation between the outside world and inside the OB is often minimal because of space and weight. Internally, noise is a problem because electronic equipment requires ventilation and fans are noisy. Heating and air conditioning noise is a problem because proper design requires space to channel the air flow and slow the velocity of the air flow.

In most audio space locating the speakers is often a compromise, particularly with the center and surround speakers and possibly even more of a compromise with the height speakers. Proper speaker placement and alignment is critical for the audio mixer to make valid decisions about levels, tone and spatial placement of the audio mix soundfield. The center speaker is a situation where misplaced speakers quickly complicate accurate monitoring for speech levels and intelligibility – the number one complaint by viewers to broadcasters.

The center and surround speakers' position is often compromised. The center speaker is often positioned too close and is not well-aligned with the other speakers, which can result in the mixer reducing the level of commentary channel. The surround channels are very difficult for proper monitoring because the space behind the sound mixer is quite minimal in many OB vans.

In one of the worst cases I've seen, the sound mixer's back was literally right up against the wall and the speakers positioned at a steep angle above the mixer's head.

An attempt to minimize room anomalies is to use nearfield monitor although room acoustics do affect the nearfields speakers and you cannot expect to avoid a room problem just by putting a speaker up closer to your face.

Position your monitor so that the front three loudspeakers can align properly. Remember how your ears work – you don't want the center-channel speaker placed above your head while the front left and right speakers are several inches lower, for example. Make sure the tweeters line up from right to left. You need to set up the speakers symmetrically, roughly at head height relative to the seated mixing position.

Free-standing and surface-mounted speakers exhibit problems depending on the distance from the ceiling and wall surfaces. These boundaries generate reflections and some frequencies will reflect off surfaces in phase, resulting in peaks in the frequency response, while other reflections are out of phase, resulting in frequency cancellation.

The problems with boundary reflections diminish as the frequency increases. Typically, when a speaker is placed too close to a surface, you will experience a bass boost. A common problem at low frequencies is the interference between the loudspeaker's direct radiation and

Figure 7.5 Bobby Carter back up against the wall: CBS OB Van

the reflection from the wall behind the speaker. At low frequencies this reflection will be delayed so much as to be in opposite phase relative to the direct sound. Depending on the relative amplitudes of the direct and reflected sounds, a cancellation dip of between 6 and 20 dB can occur in the frequency response. Addressing the low-end problems with equalization is not a substitution for acoustic treatment – there must be an acoustic solution as well.

Another important consideration for speakers is that they are properly level balanced to each other. It is critical that all of the speakers reproduce the same sound pressure level (SPL) when fed with a given reference signal or monitoring level. The goal is to get the same SPL from any channel that is given the reference noise signal, with the exception of the LFE channel. When bass-managed systems are used, this will require an audio spectrum analyzer to set both the overall level and the bass-managed level reproduced by the subwoofer.

Production Control Room

Adequate and proper monitoring in the production control room will certainly vary with the production. Production monitoring in the OB Van is often minimal because the producer and director are generally wearing headsets and are not in a position to properly monitor the sound.

Speakers

Overcoming the myriad of problems associated with inherently deficient listening spaces begins with speaker designs for critical listening. Speaker design has generally been about originating all frequencies from the same location in space and time. This clearly means that the speaker, enclosure and acoustics should not change the direction and dispersion of the sound waves. There is no doubt about the difficulties of keeping the low and high frequencies constant in the horizontal and vertical planes because of the vast difference in the size of low and high frequency sound waves.

Many speaker manufacturers have adopted the term and principles of a shaped wave guide, which physically contours the speaker and/or enclosure to guide the propagation of the sound waves. The intended results are controlling the directivity and dispersion of the sound waves in an attempt to more focus the sound at the mixer while neutralizing reflections off the adjacent surfaces including walls, floor and ceiling as well as equipment. The downside is that a tighter focus of the sound waves may also result in a narrower sweet spot.

Multi-speaker enclosures use separate transducer drivers for the high and low frequencies which can lead to uneven reproduction, phase and distortion issues.

By locating the transducers as close as possible to each other and to the listener, this positioning helps the brain aurally perceive that there is a single source point. Point source is a relatively common speaker principle because of the advantages for phase coherence, while the ultimate point source design is coaxial speakers.

Nearfield speakers, as implied by their name means the speakers are closer to the listener and because of the nature of the audio mix room in most OB the speakers are relatively small. Speaker designs seem to come in a variation of two-way, three-way, coaxial and some with porting to improve bass response, although ported speakers as well as all speakers are impacted by being too close to the walls.

Nearfield monitoring minimizes poor acoustical environments but is not a complete solution to speaker placement where the distance necessary for an accurate reference should be a consideration. Controlled directivity seems to be common with multi-speakers enclosures using separate transducer drivers with "waveguides" as well as "coaxial" speaker designs which use a single driver. Controlled directivity in both planes gives a larger sweet spot with good direct sound and coherent reflections.

Point source is a variation on coaxial speaker designs that is less exact as to where the listener location or sweet spot is because all frequencies originate from the same location in space and time. I spoke to Thomas Lund from Genelec and he told me "from a perceptual point of view organic (sound) sources above 300 Hz, like the human voice and musical instruments radiate this way naturally".

Wave guide and point source are viable designs for small enclosures so now we must consider the size of the enclosure. Most OB Vans have implemented surround sound monitoring within small rectangular spaces but above speakers for immersive sound are going to be challenging. "The Ones" speaker by Genelec can deliver an accurate listening reference at a distance as little as 50cm and an immersive impression in a 2m square – bottom-line OB size.

Virtually all manufacturers incorporate some sort of equalization and level contours to try and tame the room acoustics but no one that I could find uses a dynamic software control in each monitor like Genelec to configure, calibrate and control each monitor in the system for frequency response, level and time-of-flight or distance of the speakers from the listener.[2]

Bass Management: An Overview

Generally, small speakers require a subwoofer to reproduce the low frequencies while mixing for film and broadcast require a LFE speaker. I will suggest that this is the origin of some confusion between using the subwoofer for bass management and the role of the subwoofer as the LFE.

The overall sonic experience of live sports audio for home listeners has improved with soundbars, but I believe that live sports sound tends to have an unnatural rumble because the sound mixer cannot properly hear full frequency 5.1 surround sound. Significantly, many up-processing devices push some of the sound into the LFE channel and a lot of this content that is upmixed is definitely not LFE content. This is concerning because the overall sonic experience of live sports audio for the home listeners has improved with soundbars.

Bass management is used to create full frequency sound from smaller speaker designs. Low frequencies, below around 100 to 120 Hz, are filtered from the five main channels, routed and summed to the subwoofer. The full-range audio is routed to the bass management system where it is filtered and the low-frequency signal is sent to the subwoofer, while the remaining audio is sent on to the left and right speaker. The full-range audio signals of the center, left surround channel and the right surround channel are routed to the bass management system where it is filtered and the low-frequency signal sent to the subwoofer, while the remaining audio is sent on to the center speaker.

The low-frequency audio is then sent to the bass management module, where independent gain adjustment can be made to the overall bass levels then the low-frequency audio is sent to the bass extension or to the bass redirection.

The bass extension allows the low frequencies to be sent not only to the sub, but back to all five main channels, allowing the bass to emanate from 360 degrees. The bass extension can be disabled, in which case the low frequencies are routed to the bass redirection module only. Bass Redirection provides mutes for all 5-channel low-frequency signals. Un-muted low-frequency signals are then summed and sent to the subwoofer via the sub input.

The LFE channel provides an independent bass path that can be used at will without affecting, or being affected by, the normal bass coming from the five main channels.

You get to decide exactly when, what and how much low-frequency material is used for precisely what purpose and between the monitor system's bass management and the software bass management, you have the ability to add as much or as little bass as necessary.

Finally, a word regarding phase relationships with the Sub/LFE channel. Make sure you align your subwoofer correctly, since the sub and your five main speakers are now rather distant from one another. Remember, low frequencies are omnidirectional so you have a latitude for sub placement.

All listening rooms need to be tuned. Virtually all manufacturers incorporate some sort of equalization and level contours to try and tame the room acoustics but no one that I could find uses a dynamic software control in each monitor like Genelec to configure, calibrate and control each monitor in the system for frequency response, level and time-of-flight or distance of the speakers from the listener.[3]

The very nature of mixing in an OB Van means that the mixer is probably constantly moving their head/ears relative to the speaker (on-axis/off-axis), suggesting that a less exact listener location or sweet spot maybe desirable as presented by PATENT 535710991.

Spatial Sound Reproduction System

This sound reproduction system radiates sound from four point-source speaker arrays or columns and is particularly advantageous for close critical listening typical of the detailed listening environment of the sound mixing studio, broadcast monitoring studio or outside broadcast sound production rooms. The system may be programmed to model actual as well as virtual immersive sound formats. This speaker system may accurately reproduce immersive spatial sound produced for object- and speaker-based dimensional audio. In particular, the inventive system may reproduce ambisonics, discrete channel-based sound productions, discrete speaker-based sound productions, and soundtracks from other common immersive sound processes marketed and produced as DOLBY ATMOS®, MPEG-H, AURO 3D, DTS-X, SONY360, and others.

When properly focused and tuned, the vertical orientation of multiple sonic emitters is an ideal configuration for very small listening environments such as the mobile broadcast production audio control room where critical listening is essential. Closely spaced sound emitters, speakers and transducers form a cohesive guided sound wave which creates a frontal soundfield at ear level as well as above the listener where immersive sound really comes from. In addition to directly projecting sound from each vertical enclosure, the present system may project sound away, off-axis from the listener, to reflect sound to the sides and behind the listener.

The front speaker enclosures should be symmetrically spaced apart and located on the left and right side of a picture monitor or television screen for visual applications. The upper side-firing speakers and the lower side-firing speaker are set at about a 20° angle, comparable to DOLBY ATMOS – enabled up-firing speaker. This approach to immersive sound reproduction uses direct sound wave control dispersion, off-axis, and side-firing transducers, plus room compensating acoustics processing to create a complex dimensional soundfield. This system can self-tune to the acoustic environment using digital signal processing.

Processing and other sound wave forming and manipulating principles control sound dispersion from the multiple speakers and transducers in the vertical enclosure. In some embodiments, the inventive spatial sound reproduction system may be a network of four vertical sound (vertically oriented speaker) arrays that accurately and linearly reproduce an immersive sound experience and minimize the need for unnatural enhancement and processing to achieve spatialization. The arrays guide a balanced sound to the listener's natural ear sensitivity and work with the brain's true interpretation of sound spatialization.

Figure 7.6 shows an array according to an embodiment of the invention with six vertically oriented front-firing speakers: two upper, front-firing speakers, two middle, front-firing speakers, and two lower, front-firing speakers; and two transducers or surround, side-firing speakers to the left of the front-facing speakers.

The present invention provides a sound reproduction system using a plurality of vertical speaker enclosures, each with vertically oriented in-line transducers. The system may comprise either two or four speaker enclosures or arrays with six vertically oriented forward-facing

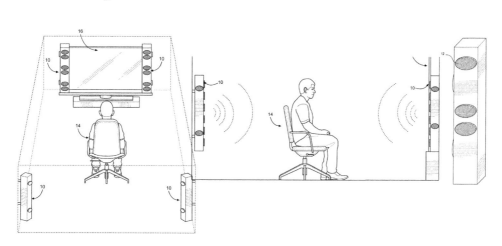

Figure 7.6 Spatial sound reproduction system

in-line transducers in each enclosure. In other words, six forward-directed speakers distribute an upper height, a center height, and a lower height. In some embodiments two side directed speakers, for left height surround and left surround, also known as side-firing speakers, may be provided for additional reflected sound reproduction behind the listener.

Figure 7.6 shows an array on each side of a monitor or television, as well as two rear speaker arrays or speaker blocks behind the listener. Each array behind the listener is shown with at least one front- firing speaker and two surround side-firing speakers. It should be understood, of course, that the foregoing relates to exemplary embodiments of the invention and that modifications may be made without departing from the spirit and scope of the invention as set forth in the following claims.

As used herein, directional terms such as upper, lower, upward, downwardly, top, left, right and the like are used in relation to the illustrative embodiments as they are depicted in the figures, such that the upward direction (or upper) being toward the top of the corresponding figures and the downward direction being toward the bottom of the corresponding figures.

The term "forward facing", as used herein, refers to components facing toward the location of the listener.[4]

Notes

1 "Samsung Home Theatre Soundbar", www.samsung.com/us/home-theater/soundbars (accessed 15.12.2021).
2 "Genelec, Broadcast & OB-Van", www.genelec.com/broadcast-ob-van (accessed 15.12.2021).
3 "Genelec Bass Management", Users/dennisbaxter/Downloads/GLM4_System_Operating_Manual.pdf (accessed 15.12.2021).
4 Spatial Sound Reproduction System Patent 535710991.

8 Mastering and Distributing Immersive Sound

James M. DeFilippis, co-author

As a broadcast sound designer, I've spent my career trying to figure out how to achieve the perfect live broadcast sound mix and distribute it to the consumer. But the fact is that no matter how good the mix is, it's going to sound different from one home viewer to the next.

Over the last few decades, the visual aspect of broadcasting has been fully explored, from black and white to color, from standard definition to 4K High Dynamic Range. Now is a chance for the television producer to fully explore the experiential value of the sound. To the sound engineer, it presents the opportunity to develop new production workflow practices and offers creative possibilities in exploring the depth of sounds that can be used to heighten the entertainment experience. For everyone involved, it provides the professional opportunity to evolve and fully embrace the creative possibilities that this personal sonic technology creates. Clearly for the savvy broadcaster who understands this challenge and delivers the amplified consumer sound experience, it provides an opportunity to differentiate their programming and retain viewer loyalty.

Future ready entertainment – Next Generation Audio can be described as integrated, immersive, interactive and intelligent. The question is how and what is the best way to get your experience and audio to your audience? How do we implement all the new features of Future ready entertainment – next gen audio? This chapter will look at mastering the audio to be distributed as either a digital stream or media file for consumer consumption.

How Did We Get Here?

Getting multichannel audio to the consumer has always been a process. Analog audio for television was carried over FM frequencies which changed after the conversion to digital television where the audio and video are encoded (compressed) and then multiplexed into a common data stream.

In the analog transmission era, Dolby™ developed a surround encoding algorithm that could be transported in a pair of stereo signals. It was problematic because the implementation by Dolby relied on analog processing that matrixed the surround mix to fit into two existing analog transmission channels. If the consumer had a Dolby Pro Logic® decoder the two channels would be de-matrixed to produce a three-channel surround format while if there was no decoder the processed audio could be presented as stereo. There was a significant technical hurdle to make the Pro Logic process signal compatible with the existing unprocessed stereo signal given the variations and processing present in analog audio transmission. NBC referred to the Dolby Pro Logic process as "super stereo" and convinced the Olympic Host Broadcaster to produce the stereo program feed using Dolby's Pro Logic in eight sports and the Opening and Closing Ceremonies at the 2004 Summer Olympics. Dolby Pro Logic never really achieved the full capability of surround audio production, and while it worked within

DOI: 10.4324/9781003052876-8

limitations of analog transmission by 2008 there were alternative solutions to multichannel production and distribution.

Future Ready Entertainment: Next Generation Audio (NGA)

Digital audio provided new opportunities for alternative methods to the analog approaches in order to provide advance audio features and programs to the consumer. Standards were established to allow for practical exchange of media streaming and files between consumers and producers. These innovations did not come for free, requiring both a re-tooling of the production process as well as paying license fees paid by device manufacturers to use these new technologies. Today digital audio is either delivered as a physical disc or a downloaded file, transported over terrestrial, cable and satellite networks using format standards so the consumer's equipment – computer, smart TV and handheld devices can receive the advance audio and video content, decode the digital stream to audio and video for consumption.

Compatibility between audio programs requires a unified set of production standards; there are many organizations that develop technical and production standards such as the International Telecommunications Union (ITU), Audio Engineering Society (AES), International Organization for Standardization/International Electrotechnical Commission (ISO/IEC), European Telecommunications Standards Institute (ETSI), Society of Motion Picture and Television Engineers (SMPTE) and the European Broadcast Union (EBU) are some of the principle organizations that develop standards. However, problems may occur with compatibility between different productions when different organizations have different standards for the same thing or they are not understood by the practitioner.

There are several variations of digital television coding standards that align along territorial and regional boundaries. Digital Video Broadcasting (DVB), Advanced Television Standards Committee (ATSC), Integrated Services Digital Broadcasting (ISDB), Digital Terrestrial Multimedia Broadcast (DTMB) and Digital Multimedia Broadcasting (DMB) developed comprehensive transmission standards that transport compressed digital audio and video signals. Each of these transmission standards have their own requirements and technologies as set forth in their respective standards. The data capacity of these transport technology methods is determined by the available channel bandwidth, modulation coding, and data encoding including forward error correction (FEC). Receivers successfully recover these digital signals as long as the received signal strength is above a threshold value (called 'Eb/No'). The recovered audio and video quality is not affected no matter how far away the receiver is from the transmitter, unlike analog transmissions in which the audio and video quality diminishes with the loss of received signal strength. The video and audio compression quality is dependent on the available data capacity. In general, there is a minimum data rate threshold for a level of video and audio quality. For example, the ATSC adopted several compression methods for video and audio to allow versatility in the use of the available bandwidth and with the use of existing IP standards the audio and video can be streamed over the air, cable, broadband and mobile. The video codec, HEVC (H.265) can support video coding bitrates from 100kb/s up to 30Mb/s, while the audio codecs, MPEG-H and AC-4, can provide different modalities and qualities using bitrates from 35kb/s to 1Mb/s.

2015 brought the introduction of a new generation of audio resources from AURO-3D, Dolby ATMOS, MPEG-H 3D and DTS X that can be coded and embedded into the transmission stream. ATSC 3.0 adopted two audio formats AC-4 and MPEG-H 3D to support the ATSC 3.0 standard. Fraunhofer is the developer of MPEG-H 3D while Dolby developed AC-4. Even with the most extensive production (7.1+4) formats and extreme quality, using these NGA codecs, audio programs can be encoded, distributed and decoded by the consumer device and rendered as long as it is processed with the appropriate digital codec. However, these audio codec decoders are not cross-compatible and consumer devices need a license for each.

Dolby Atmos© is an immersive sound format, while AC-4 compresses the immersive audio program to the required bitrate for optimum delivery across all devices. Digital television transmission requires a high transmission channel bitrate for the increased channel count and features. The bitrate determines the depth of the audio channels, fidelity and capabilities of each channel. Bitrate standards for audio formats usually include the data rates for immersive, interactive and intelligent/personalized features as well as lower bitrates for lower resolution requirements such as mono, stereo and surround. Additionally, AC-4 and MPEG-H 3D support dialogue enhancement, loudness, dynamic range, multiple languages and personalization with audio bitrates between 30kbs to 1Mbs. While both MPEG-H 3D and AC-4 have the ability to synchronize with the video signal, AC-4 is encoded within the same time interval as the associated video frame time, providing a simpler transition (cut/fade) between programs. By contrast, compressed video requires bitrates between 1 Mbps and 35Mb/s; however, the video and audio, and ancillary data, have to fit within the data bandwidth available in the 6/7/8 MHz OTA RF channel.

Microphones are brought into the mixing console and all sound elements are combined into an immersive mix. Immersive audio authoring – channel-based audio groups are created – 5.1.2, 5.1.4, 7.1.4 and authored with metadata. Immersive audio monitoring renders speaker feeds to audio mixing space for mixing. Immersive audio channels encoded as AC-4+JOC and delivered as an immersive sound bit stream with metadata over MADI or SDI to contribution encoder.

Legacy audio delivery – Stereo or 5.1 can be created in the mixing console and embedded for delivery to contribution.

Digital coding technology facilitated the change from analog to digital audio and led to the use of digital data compression and file formats, which are described in all global standards for audio and video exchange. Why are there competing standards? Profits – the consumer manufacturers pay a license fee for each format decoder included in their electronic device. Samsung, Sony and LG were leading adopters of ATSC 3.0 and have included the new standard in their devices.

There is tremendous demand by consumers for media content, thus generating robust competition between electronic device manufacturers, software streaming companies and media content owners (film studios, television networks, recording companies), leading to a plethora of solutions, providing choice to both the distributor and consumer. In the case of ATSC 3.0, two audio technologies were evaluated and standardized, Dolby's AC-4 and

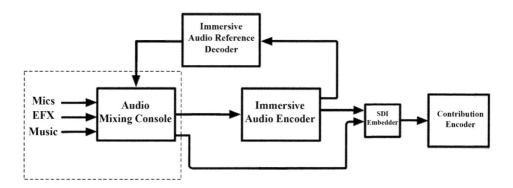

Figure 8.1 Immersive audio encoding signal flow

Fraunhofer's MPEG-H 3D audio. LG, Samsung, Sony and others have licensed both codecs depending on the market (AC-4 for US use, MPEG-H for Korea).

To prepare the audio for distribution either by file, over the internet or transmitted linearly over terrestrial transmitters, satellite transponders, or cable TV systems, it must first be mastered using authoring tools from one or more of the authoring formats – Dolby® Atmos™, MPEG-H, Auro 3D and ambisonics rendering. Digital audio is mastered and sent to the consumer to be rendered to the appropriate listening configuration. Mastering for immersive digital audio includes bit rate reduction, localization tools and metadata authoring. Dolby Atmos dominates film production with some competition from Auro 3D, however MPEG-H seems well suited for internet, broadcast and gaming applications due to the ease of personalization (interactive controls) for consumers.

Metadata allows the audio to be correctly handled and processed throughout the production and transmission ecosystem. The European Broadcast Union (EBU) developed Audio Definition Model (ADM) to give a complete technical description of the audio within the file but is not intended to give instruction about how to render the audio. The ADM consists of information about the audio track, stream and channel format, group format about what channels belong together as well as information about the audio within an object. ADM has been formally standardized by the ITU-R as BS. 2075-2.[1]

Immersive metadata includes parameters for loudness, dynamic range control (DRC), channel format, object name, object positional coordinate (xyz), range of loudness controls as well as interactive personalization features such as dialog enhancement, choice of announcers, etc. Dynamic sound elements require dynamic metadata with time stamps, rendering mode, size of object plus information about width, gain, diffusion and divergence.

In the US, Canada, Mexico and South Korea terrestrial transmission standards for television is developed by the Advanced Television Standards Committee (ATSC). ATSC 1.0 was adopted by the US government, mandated and ushered in the transition from analog TV to digital TV, including HDTV and Dolby AC3. In 2017, the ATSC finalized the specification for a new terrestrial television standard called 3.0 which is not mandatory in the US but was adopted by South Korea in time for deployment prior to the 2018 Olympics in PyeongChang. ATSC 3.0 adopted Dolby AC-4 and MPEG-H as part of the ATSC 3.0 standard for immersive audio that supports full decoding and rendering by consumer devices – the television, set-top box and handheld devices that support ATSC 3.0. Note – ATSC 1.0 specified Dolby AC-3 as the digital audio format, which in my opinion further promoted the surround sound standards that consumer electronic manufacturers adopted.

In other parts of the world, there are other digital television systems. DVB-T – Digital Video Broadcasting – is a European-based standard for digital television, adopted in many countries including in Europe, Africa, Mid-East, Asia and Australia. DVB is currently evaluating an update to the current standard (T2) to include immersive audio codecs. Integrated Services Digital Broadcasting (ISDB-T) Japanese standard for digital television and radio used in Japan, Philippines, Brazil and other South America countries. Japan is launching their update to ISDB-T, including ultra high definition video (8k) with 22.2 multichannel audio using AAC audio coding. DTMB Digital Terrestrial Multimedia Broadcasting developed in the People's Republic of China. DTMB is in use in China as well as some countries in Asia, the Caribbean and Africa. China is evaluating the inclusion of immersive audio in the next generation of their standard.

Dolby Atmos

Dolby Atmos is not one specific codec but is a philosophy of immersive audio sound production and presentation, for many applications such as theatrical, home video, broadcast, streaming and mobile. It can be carried as an extension to existing codecs like E–AC3 or in a new codec, AC-4 which has been specified in ATSC 3.0 and adopted by US broadcasters. Currently support for Atmos in televisions, AVRs and soundbars uses E–AC3 w/JoC (Joint Object Coding). E–AC3, also known as Dolby Digital Plus© Dolby Atmos uses sampling rates up to of 32kHz and maximum bit rate of 6.14 Mb/s to reproduce up to 7.1 surround channels plus up to four height channels. Dolby Digital Plus with Atmos, is backward compatible with non–Atmos aware decoders.

MPEG-H

MPEG-H is not just an audio codec. MPEG-H allows for immersive, interactive and personalized audio to be produced live or via a plug-in to a digital audio workstation (DAW). As a plug-in the audio is enhanced, then rendered in the DAW into MPEG-H, including both static and dynamic metadata. For live encoding, the audio is sent from the mixing desk to a MPEG-H mastering unit such as Junger Audio MMA for assigning enhanced features to each channel and generating the accompanying metadata. After the audio mastering and enhancement, the audio and metadata is delivered to the MPEG-H encoder using a unique transmission format and Control Track©, which conveys the metadata as PCM audio signal.

Digital audio compression has continuously improved since the introduction of MP3 and AC3. While digital video compression can trade off image size, bit rate, frame rate, digital audio compression has to employ a variety of techniques (some are really tricks) to reduce full uncompressed audio data from 3Mb/s (stereo PCM) to as low as 30kb/s. For more complex audio programs, additional tools such as spectral mirroring, joint object coding, high order ambisonics are able to encode a 7.1+4 immersive audio program (16 PCM channels at 24Mb/s) to a data range of 600kb/s to 1Mb/s. While the Next Generation Audio (NGA) codecs use many of the same tools to compress these high channel counts, they are not cross-compatible in terms of decoders. The ATSC 3.0 supports two different audio formats, Dolby AC-4 and

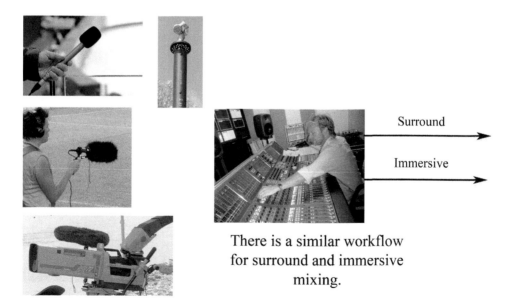

Figure 8.2 A similar workflow exists for both surround and immersive mixes

MPEG-H, which have about the same compression efficiency but are different in the compression tools used, the metadata format, and the methods used to describe immersive audio objects.

When Dolby introduced Atmos for cinema sound in 2012, Dolby expanded the then horizontal surround sound reproduction (5.1, 7.1, 9.1), added overhead channels/speakers and audio objects. The overhead dimension was clearly lacking in the original surround sound equation. Atmos provides the ability to isolate sounds, not in a speaker channel but in acoustic space. These audio objects can be static or dynamic and dynamic in position, time and amplitude. Significantly object-based technology allows for the audio elements to be kept in the bitstream and in a session allows for additional positional data to be added on top that gets rendered with the playback independently of the playback system.

Although most people today enjoy listening in stereo over TV speakers, Atmos provided a unique theatrical experience over many different devices that people can enjoy. It gave a new dimension to movie sounds and the consumer experience.

While a movie theatre is a controlled space, with known acoustics, speaker placement and volume, in order to bring immersive sound to the home, the Future Ready NGA codecs need to be adaptable to the variations in the home environment. Thus, Future Ready NGA decoders include a render engine. The renderer can take the full immersive program, including audio objects, and present them to the viewer in the best possible way, given the limitations of the home environment. In addition, this rendering can also be used to personalize the audio program with features such as dialog enhancement, choice of language or announcer, and special audio services for the vision impaired.

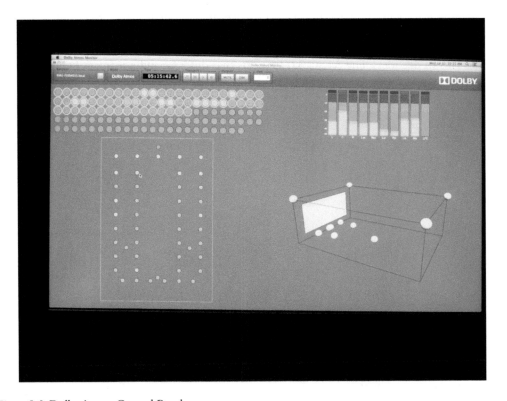

Figure 8.3 Dolby Atmos Control Panel

Production and Distribution

Today there are immersive sound production and distribution authoring tools that enable the production of immersive, interactive and intelligent audio programs, ready for consumer enjoyment. When Dolby first proposed immersive sound to the film industry in 2012 there were no immersive sound mixing consoles or operators with conceptual knowledge or experience. Dolby not only developed the technology and workflow to produce and monitor immersive sound but put all their resources behind getting Atmos into the film mixing houses and stages.

Producing Atmos for cinematic release is very different than producing for live events even though many production principles remain the same. All productions start with two paths for the production of audio – beds and objects. Beds are channel-based mixes and objects are discrete audio elements. Beds are associated with fixed speaker configurations that have defined channels. A bed is similar to mixing a channel-based format like surround sound. In film audio production the bed is often used for dialog, music and backgrounds. The bed is not rendered like objects and is the foundation for all mix variations. Remember: every Dolby Atmos mix starts with a bed, even as small a stereo.

Objects are unmixed mono or discrete stereo sounds that go to the Atmos renderer along with appropriate metadata (spatial, temporal, amplitude). The metadata informs the renderer where in the sound field that sound object should be placed, within the room configuration, so the renderer requires the knowledge of the speakers available and their relative locations within the listening space, in order to provide the listener with an accurate and realistic reproduction.

In live television production, the bed is often the immersive sound channel configuration – left, right, center, LFE, LS, RS, LFH, RFH, LSH and RSH. Dolby Atmos production must have a bed, stereo, 5.1 or 5.1.4 and cannot be produced live strictly with objects.

Some live television productions have used E-AC3 to produce an Atmos immersive audio program. In this type of production, a base surround program (usually 5.1) is mixed and distributed to the regular distribution channels. A 'height' ambience mix is created and together with the base surround mix, are sent to a E-AC3 encoder enabled for Atmos. Typically there is no dynamic objects or metadata so the encoder sets fixed parameters for the Atmos enabled decoder. This approach of using E-AC3 for Atmos is backward compatible, any E-AC3 decoder can decode the base surround mix with the height information presented in the surround mix. This is accomplished through the use of Joint Object Coding (JoC), a feature of the Atmos E-AC3 encoding.

Dolby Atmos was designed for the cinema. Their unique cinema processor determines which array of front, back and side speakers best recreates the sound experience in the theatre. For example, Dolby Atmos in theatres uses a 10 channel 9.1 (7.1.2) bed for ambiance stems and center dialog leaving 118 tracks for objects. Dolby Atmos for the Cinema uses up to 128 channels with spatial audio description metadata for every channel. Dolby Cinema applications can give each speaker its own unique feed based on the speakers' exact location enabling precise panning and localization for overhead sounds such as birds and planes.

This technology which was initially created by Dolby for commercial cinema applications was later adapted for home cinema. Dolby Atmos Home uses a conventional 5.1 or 7.1 surround speaker layout with two to four overhead speakers. Dolby enabled up-firing speakers (see Chapter 7, Monitoring) using a maximum of 12 channels for reproduction. Atmos for home cinema usually reproduces with one 10 channel (5.1.4) bed channel, an LFE and ten dynamic objects. There is much disagreement over whether E-AC3 Atmos supports dynamic objects, however we know there is no dynamic metadata in the E-AC3 specification that can define a 3D sound object.

Dolby Atmos treats specific sounds as individual entities – what Dolby calls audio objects. Dolby defines an object as an audio channel that includes metadata. Objects can be placed in space in precise locations and moved anywhere in the three-dimensional space. For example, if the object is panned from back to front or bottom to top the sound will pan through each real and virtual speaker in its path, using the location and pan automation data and finally rendered to a smooth movement of sounds.

Dolby Atmos production requires the hardware and software components to monitor, create and playback Dolby Atmos content. The Dolby Atmos renderer is resident in the Dolby RMU hardware as well as in plug-ins for ProTools©, Nuendo© and other DAW platforms. Dolby Atmos Mastering Suite and Dolby Atmos Production Suite include the Dolby Atmos renderer where the immersive audio mix is routed and processed through the production suite plug-in. Dolby's production suite runs on the same system as ProTools, and can be demanding on CPU power.

Cinematic Production vs. Live

All cinema mixing facilities that produce Dolby Atmos have specific requirements for certification and many large mixing stages use a standalone Dolby processing unit called a Rendering Mastering Unit (RMU), which Dolby provides to film mixing facilities and large productions. The RMU hardware uses Dolby's Cinema Mastering software with MADI or Dante audio interfaces to connect directly from the mixing desk or DAW to the RMU. The RMU supports 128 channels in and 64 output channels plus bass management on the surround channels. The RMU is not dependent on a CPU host as it runs on its own hardware. The mixing room can be calibrated for its unique number of speakers, distance between speakers, number of subs and speaker model. The Dolby Atmos renderer resides in the RMU and provides 3D panning tools to place the objects, monitoring controls, metering controls for peak levels and loudness, room equalization (EQ) to set bass managed speakers, B chain processor for EQ, dynamics (AGC and leveling) and delay plus mastering controls for file delivery to the encoding. Dolby Atmos monitor application enables communications with the RMU and performs operations by screen and mouse.

I first encountered Dolby Atmos in 2014 at the Technicolor mixing stage in Hollywood. The mixing console was a modified Euphonics with over 96 mix channels. There was a MADI interface to the Dolby RMU where Dolby Atmos operations were monitored on screen. The RMU controlled the sound monitoring in the theatre size room using the immersive 11.1 format – 7.1.4. In post-production applications such as film, drama and entertainment it is likely that there are a multiple of deliverables from stereo through 7.1.4. A significant advantage for new advance audio formats is the ability to render/re-render additional mixes. The separate stems can be constructed in the RMU and be configured in the re-render outputs. When a film is mixed in 7.1 channel based all the sound elements are baked in – you can't go back and pull an ingredient out without a remix. With objects you can adjust any sound element in the mix including removing the element entirely.

MPEG-H 3D

The MPEG-H 3D audio system was developed by Fraunhofer and is based on the ISO/IEC Moving Picture Experts Group (MPEG) standards to support coding audio as audio channels, audio objects, or in a novel approach as higher order ambisonics (HOA).[2]

Fraunhofer's audio system technology is designed for superior delivery of content while providing a differentiating edge to broadcasters who can now deliver an immersive, interactive and intelligent experience to the consumer regardless of platform. MPEG-H 3D may be used to deliver/master immersive sound as channels, objects, and HOA components as well as transport mono and stereo objects. What Fraunhofer has figured out is a good sounding bit-rate reduction codec that gives the audio producer additional full-fidelity controllable audio channels with which to design a more engaging entertainment experience.

While earlier generations of audio were primarily intended to deliver a variety of enhanced listening experiences, MPEG-H Audio is more than a passive listening experience, it can also offer user-controlled audio channels. Up to 15 separate outputs can be used for immersive sound channels, interactive channels, multiple languages – MPEG-H offers complete personalization and control. MPEG-H 3D Audio can support up to 64 loudspeaker channels and 128 codec core channels similar to Dolby.

Interactivity allows the audio producer, while delivering a sports audio mix for over-the-air broadcast, an option to offer an interactive or personalized mix for the discerning and passionate viewer. Interactive audio permits the listener/consumer to select the audio elements (components) that they are interested in during broadcast. From the contestants on your favorite reality show to a variety of commentary or translations, personalized audio offers interactivity in consumer choice.

Interactivity can go beyond the ability to select a different language. As a sound producer, I may want to use the audio from the athletes and coaches, but in reality this can be dangerous because of a concern over transmitting offensive language. The colorful language is a problem because there are restrictions on over the air broadcasters with the language that can be transmitted. With interactive channels, I can put the coach's audio on a separate channel that the user can choose to hear or not. Additionally, these interactive channels can be monetized through sponsorships or fees. If you are a fan of a sport, of a coach or an athlete, you might be willing to pay to hear this audio and be willing to accept the language.

MPEG-H was designed for hybrid delivery over both over-the-air (OTA) and over the top (OTT), also known as streaming.

Dynamic Objects

As of 2021 most live immersive sound production used little or no dynamic panning or extensive cinematic type of production, although as the audio practitioners gain experience the immersive sound production will increase in complexity. Immersive sound authoring from MPEG-H offers the possibilities of live dynamic panning which I have used and found very effective for motorsports, extreme skateboarding and snowboarding competitions. See Chapter 10.

With the goal of creating heightened sound realism of the immersive experience for events such as Half Pipe snowboarding and Big Air skateboarding, the illusion of vertical sound was created using the upper audio channels. The result? When the viewers see the athlete on the snowboard go airborne, they also hear the sound of the athlete over their head as well as horizontally on their left and right. The outcome is a new sense of immediacy and reality for the consumer and a new opportunity for broadcasters and networks to broaden their consumer base.

I am involved in live sports broadcasting and can readily see how the personal sonic technology that Fraunhofer has developed could enhance the consumer experience. Motorsports comes to mind because of the vast and rich sound sources available at the races. When viewers are watching and hearing vehicles racing at 200 mph, they would be able to mix in a favorite driver in his car conversing over the crew radio, some track sounds, and enjoy it all in a variety of room sizes and speaker modes with the touch of a screen.

An extremely useful feature of MPEG-H 3D Audio is the ability to rebalance the program. This feature permits the user to adjust the volume of the program track and the volume of the commentator track independently of each other. Everyone has experienced a program where it was difficult to understand the dialogue – either listening in a noisy environment or the results of a bad mixing process. By simply turning the voice track up the speech intelligibility can be significantly improved. Additionally, the individual voice tracks can be optimized for the specific listener with equalization if necessary. For example, there are many individuals with hearing issues and one of the features of MPEG-H 3D Audio is loudness compensation which is a useful feature to modify the dialog level and yet maintain the overall balance of the audio mix. This feature can be adjusted for the individual viewer.

Different Viewing Platforms

Broadcasters are facing stiff competition in retaining viewing audiences who are turning to alternative content sources and are choosing to consume content on many different platforms – from traditional televisions to tablets to smartphones – many of which offer a more convenient, interactive and personalized experience. Additionally, the consumer's viewing environment as well as their personal preferences can easily be optimized and played out. Consumers would be able to contour the sound for speakers in the kitchen, bedroom, living room, home theatre, as well as for ear buds for listening over the phone and tablet. The fact that media is consumed in a wide variety of conditions validates enhanced and personalized audio components, particularly for improving dynamic range and intelligibility. The use of audio objects allows for interactivity or personalization of a program by adjusting the gain or position of the objects during rendering in the MPEG-H decoder as well as binaural rendering of sound for headphone listening.

The sound control can vary with the user and could be as advanced and interactive as desired or significantly simplified by using preset buttons which can be preprogrammed to do basic functions such as boost the dialog or limit the dynamic range. For example, in a noisy environment when an individual user might need to hear the dialog above the background noise, he or she simply clicks dialog boost preset. Interestingly, presets can offer a creative zone for alternative mixes or objects, all of which can be controlled with parameters determined by the sound designer or show producers.

Controlling the broadcast sound reproduction quality across so many platforms is a daunting task and for the less sophisticated user, all this flexibility comes with a reset button that would default to the production sound mix. Significantly is the fact that individual aural health and subjective taste are clear arguments for personalization and will be a differentiator in entertainment delivery. Over the last decade, there has been tremendous improvement with audio reproduction – now people can hear good audio. This trend will continue into the future with soundbars, enabled by advanced DSPs and wireless interfaces.

For those that want user control, MPEG-H 3D provides user management for personalization and ancillary audio elements specific to the production such as additional languages and commentary, even audio from the coaches or radios.

But MPEG-H 3D is more than just immersive and interactive. It is a bandwidth efficient agile format that not only can deliver immersive sound, but also surround and stereo with interactive elements allowing for a complete personalized audio experience for the consumer. Objects may be used alone or in combination with specific audio channels or within a complete HOA mix thus enabling MPEG-H Audio to become a desirable format in which to produce audio programs.

The New Paradigm: Rendering

Rendering is converting a set of audio signals with their associated metadata to a consumer's preferred configuration of audio speaker feeds and their arrangement or for headphone listening.

How does this all work within the broadcast environment? Let's begin with the ability to deliver several program audio mixes – immersive, surround and stereo. Many audio practitioners would agree that delivering one immersive mix is difficult enough with only one mixer and one mixing desk. With MPEG-H, the audio practitioner does not have to deliver more than one mix; instead they can produce at the highest standard and format – immersive – and with the use of intelligent metadata, let the consumer devices render the appropriate mix to match their listening speaker or headphone setup.

This concept is not new because currently sound mixers deliver a good surround sound mix that is downmixed in the "set top box" in the home. The difference is that pre-ATSC 3.0 set top boxes did not render, but combined channels in a fixed downmix ratio. This process essentially adds the left, center, right, left surround and right surround components together to generate a stereo audio mix. The problem with any summation type downmix solution (adding and subtracting audio levels) is that you always compromise between the stereo, surround sound and immersive mix because you are adding the sound from the channels together to get the derivative mix. This certainly would be a greater problem with downmixing from immersive to surround, or even worse, from immersive to stereo because each combined mix gets more sound added or subtracted on, thus diluting the details in the mix.

Now consider an audio production where all the audio components and parameters exist in RAM and the user interface intelligently outputs stereo, surround, and immersive sound plus interactive audio and proper listening output according to the user's speaker setup or listening device tailored to the individual listener tastes. Metadata directs the set top box (or listening device) how to render the audio channels according to the consumer's reproduction format all while maintaining the audio producer's creative vision.

The MPEG-H 3D Audio decoder renders the bit stream to a number of standard speaker configurations as well as the ability to adjust for non-standard speaker setups. Significantly, MPEG-H Audio includes a "scene-based" rendering method that intelligently interprets the number of audio channels to the particular consumer device, be it phones, tablets, over-the-air broadcast and over-the-top broadband.

The Fraunhofer system is easily implemented by both broadcasters and consumers. It is internet-ready for a great listening experience on every device. Building on Fraunhofer's experience developing HE-AAC, the native surround audio codec of iOS and Android, MPEG H is part of the DASH specifications and thus supports stutter-free streaming and audio IPF-frames for easy DASH bit stream switching and splicing for ad insertion. It includes a multi-platform loudness control to provide a tailored experience for a viewer's device and listening environment. On the production side, it also includes the HOA technology, which is backward-compatible with the systems and practices used today for AC-3 or HE-AAC surround sound broadcasting, and offers a staged approach to implementing new features.

Any or all of the MPEG-H 3D Audio interactive features can be deployed by a broadcaster or by a content producer allowing them to control the presentation of the audio program and be responsive to the interests of their listeners. Fraunhofer has outlined an implementation plan as a four-stage approach. Start with improved coding efficiency providing for transmission of today's surround sound at 50 percent of the bit rate due to improved coding efficiencies. The addition of interactive objects seems to be an easy marketing feature in an age of games and apps. Enhanced 3D sound with height channels is necessary to complete the illusion of immersion.

With MPEG-H 3D Audio the features can be implemented as desired or needed and at the pace of the individual broadcaster. For example, MPEG-H 3D Audio has personalization features that provide the viewer with the interactive ability to select different languages; to rebalance the audio elements of a mix; to program alternative audio channels; and to embed merchandising and value-added features into the audio datastream, plus more.

Clearly MPEG-H 3D Audio brings a new level of content production options, most notably enough discrete channels to create and support a range of immersive sound formats.

There are substantial considerations for a broadcaster contemplating a transition to the MPEG-H 3D Audio format because of the wide variety of options and features but also due to the availability of several immersive sound formats with different and not particularly compatible configurations.

Ambisonics

Ambisonics treats sound equally from all directions and all channels are used for voice, ambiance, effects and dimensional productions instead of an approach that puts the main sources of sound through the speakers in front of the listener. Ambisonic works within Atmos and MPEG-H as an additional coding to accommodate the additional channels.

Higher order ambisonics as a distribution solution has always intrigued me because it offers some interesting playback possibilities due to the capability of rendering to any particular speaker configuration, from stereo to immersive formats up to 22.2 and more. This capability has interesting possibilities when trying to optimize the sound for a wide variety of soundbar manufacturers. Ambisonics was adopted by Google as the audio format of choice because of the feature of accurate and realistically tracking audio to the picture. See Chapter 4 for more on ambisonics.

Live Emissions

As previously stated, Dolby AC-4 (Atmos) and Fraunhofer's MPEG-3D audio are not cross-compatible; thus, to deliver live the signals a baseband conversion needs to be monitored and adjusted prior to outputting in AC-4. MPEG-H and Dolby Atmos can be mastered on the Junger Audio, Linear Acoustics or similar mastering unit into a digital file or stream.

Junger Audio

To produce live MPEG-3D audio an audio mixer needs to use a MPEG-H authoring tool such as the Junger Audio MMA – monitoring, mastering and authoring unit with 3D panning tool, monitoring controls, metering controls for peak and loudness and set parameters for the interactive features.

Junger Audio, a Berlin-based developer and manufacturer released its Multi Channel Monitoring and Authoring (MMA) system that has a robust set of tools and is based on the Fraunhofer MPEG-H format. The MMA is a set of tools to create dynamic as well static audio channels and objects, 3D panning with dynamic object metadata supporting output formats from mono to 7.1+4H channels. Junger Audio and Linear Acoustics not only meter the audio in real time, but can apply loudness adjustments based on an adaptive wide band loudness control plus dynamic range control and processing.

The Junger device generates the metadata for MPEG-H authoring, which includes DRC, loudness controls, static positioning (3D panning), dynamic panning, channel organization as well as parameters for the object channels.

The MMA supports 64 channels in and 64 output channels plus bass management on the surround channels. A monitor application enables communications with the MMA and perform operation by screen and mouse and the RMU is not dependent on a CPU host. The Junger Audio MMA output is PCM or analog so take note that this unit, as of publication, does not code directly to AC-4; that is accomplished downstream in another coding device like the Linear Acoustics series of authoring devices. MPEG-H 3D audio license is resident

Figure 8.4 The Junger 3D panning and monitoring. The Junger device generates the metadata for MPEG-H authoring which include DRC, loudness controls, static positioning (3D panning), dynamic panning, channel organization as well as parameters for the object channels

in the Junger Audio MMA core processor for authoring, monitoring and rendering. The Multichannel Monitoring and Authoring (MMA) can work with up to 15 audio channels and 1 metadata track. Loudness monitoring and metering, leveling controls, output emulation for downmixes, loudness and DRC controls, output formats from mono to 7.1.4, audio listening controls, interactive features parameter set as well as metadata authoring and monitoring.

Since loudness management is part of compliance and is monitored, these authoring devices have built in audio leveling algorithms. Level Magic is Junger Audio's loudness algorithm that implements a wide band adaptive control that works without filtering and band separation. This algorithm not only controls the overall loudness but can also adjust the independent levels for all input channels and objects.

While PCM audio flows through the box, it is processed, while metadata is authored and sent for transmission encoding via the MPEG-H Control Track signal. I/Os for the Junger device are AES, MADI, SDI and 12 channels of analog output.

When the end of the chain is intended to be encoded in AC-4, a suitable encoder such as Linear Acoustics LA-5300 encodes 16 channels of PCM audio to a Dolby AC-4 bitstream and can transcode Dolby Digital and Dolby Digital Plus to AC-4 along with real-time loudness leveler when encoding to AC-4.

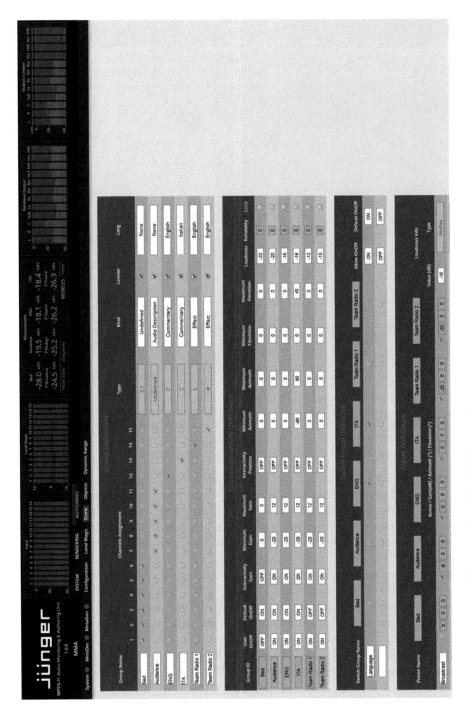

Figure 8.5 Junger metadata control panel and 3D room view

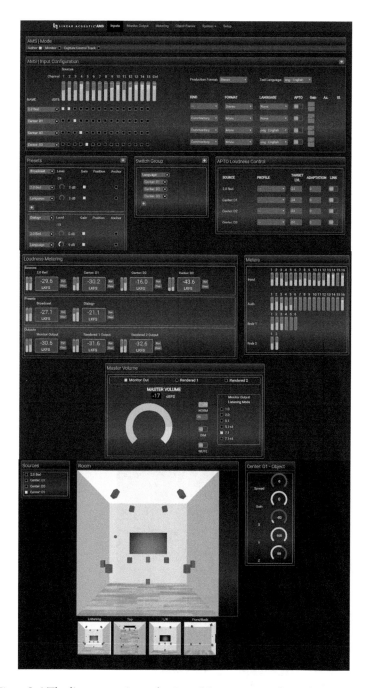

Figure 8.6 The liner acoustics authoring, object panner and monitoring system – AMS

The Linear Acoustic LA-5291 Audio Encoder encodes PCM to Dolby Digital Plus and Dolby Atmos (via Dolby Digital Plus JOC) for production workflows, platforms, and delivery streams that do not require Dolby AC-4. The LA-5291 offers decoding, encoding, and trans-coding to and from PCM and select Dolby® coded formats for up to 16 audio channels. The LA-5291 performs similar functions as the Dolby BP591.

The Linear Acoustics – AMS Authoring and Monitoring System – is a comprehensive real-time authoring and monitoring system for ATSC 3.0 Digital Television Systems.

The AMS meters the loudness parameters on all 15 channels as well as provides numerical data of short-term, momentary and integrated loudness measurements. The AMS authors the MPEG-H audio stream including the assignment of the 15 channels to 10 channel groups, identifying objects and their interactivity controls, object positioning, DRC and even some upmixing controls for mono or stereo to 5.1 or 7.1.

Linear Acoustics AMS – Authoring and Management System uses a web interface for real time authoring, rendering and monitoring advance audio content for ATSC 3.0 digital television. AMS simultaneously delivers advance audio features for ATSC 3.0 and 5.1/2 channel audio for ATSC 1.0. AMS does input/output routing configuration for up to 36 channels including 15 channels for authoring plus 1 control channel and up to 12 channels for monitoring, audio mixing and object panning, interactivity controls, DRC, upmixing, loudness processing plus real-time loudness and level metering.

Up-producing audio signals is discussed in Chapter 4 and many solutions are plug-in based where the live audio signals are bussed from the mixing desk to a computer platform that runs the plug-in and the processed audio is bussed back to the mixing console.

Auro 3D

AURO 3D is DAW Plug-in based – AAX for Macs. Even though, in a live application, Dolby Atmos and MPEG-H 3D work internally in PCM audio, the mastered audio is outputted in a compressed data for distribution and transmission. Auro 3D is unique because it is its own platform, production tools, and codec that delivers native discrete 3D content to a proprietary renderer that is not compatible with any other immersive sound format.

Auro 3D is a unique workflow with PCM audio using non-compressed, non-lossey audio from beginning to end. This creates some interesting backward compatibilities with some

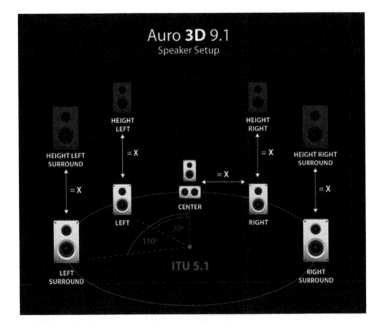

Figure 8.7 Auro 3D speaker setup

formats as is the case between ATSC 1.0 and ATSC 3.0 which use incompatible audio codecs and formats.

Auro 3D is a variation on 5.1.4 with the addition of the VOG (Voice of God) overhead speaker. Auro 3D uses a speaker layout that constructs three vertical layers. Layer 1 Ear level Surround, Layer 2 height and layer 3 top/overhead. The VOG speaker is unique to AURO 3D immersive sound format. The top layer over head is directly over the listener and offers unique production possibilities and mixing techniques with highly directional off axis sound sources.

Scalable from 5.1 to 24 channels with 100 percent channels separation, low latency – no object based data. Single file distribution for 2D and 3D, compatibility with Blue Ray, HDMI, DCP. 3D simulation in headphones using binaural technology. Upmixer plug-in – Auro 3D engine upmixing algorithm converts legacy mono, stereo and surround to Auro 3D format.

DTS-X

DTS-X is another competing immersive sound mastering codec that is included in many home AVRs along with Dolby Atmos. DTS-X is a 24bit lossless codec that supports high channel count and object-based immersive audio over uncompressed PCM audio with embedded metadata. DTS-X uses an open platform object-based immersive audio solution called Multi Dimensional Audio (MDA), which gives content creators control over placement, movement and volume of sound objects. MDA Creator allows mixers to map the sound signals into 3D space creating the spatial information of each sound element as metadata. Since it is mapped in 3D space it is dependent on the DSP to place the object in the room regardless of how many speakers and is compatible with legacy and future configurations.[3]

DTS-X is not channel based and not restricted to a specific speaker layout used to represent a multi-dimensional audio space. For example, DTS supports non-height formats such as 5.1 and 7.1 as well as height formats up to 11.2 channel systems using as any as 32 different cinematic speaker locations in the home theatre and virtually unlimited number of speakers in a theatre installation. DTS Speaker remapping engine can calibrate itself to the speaker configuration of the room.

DTS, as well as Dolby Atmos, use bed tracks which are the foundation channels for audio objects to be layered on. Bed channels are static, however you can pan through the bed channels. Backgrounds are usually basic ambiances such as wind, birds, and traffic, typically anything outside without specific sync. Audio objects are specific sync sounds such as airplanes, cars, footsteps and even voices. DTS-X allows the user to adjust the volume of individual object channels. For example, the dialog can be adjusted without changing the entire center speaker.

DTS-X takes the audio and positional data and renders it into any speaker system including headphones, 5.1 or 22.2. DTS-X Creator is complete and creates an immersive sound monitoring setup using headphones or through pro tools. DTS Headphone X Monitor on any headset and are backwards compatible with legacy DTS-HD. However, DTS 3D Decoder and Dolby Atmos are not compatible.

DTS-X Media player enables frame accurate Qc-ing of DT-X and legacy DTS-HD, DTS Digital Surround down to AAC-LC audio streams. DTS-X Encoder Suite delivers the ability to create, modify and QC bitstreams with up to 12 (11.1) audio channels with objects and metadata for encoding to Blu-ray, Ultra HD and other premium digital media formats. Additionally, DTS-X can modify legacy DTS-HD streams.

Sports broadcasting is a perfect application for personal sonic experience and technology of Next Generation Audio. Live production begins in the control room or OB Van where the immersive sound production and creative decisions are accomplished over a speaker or channel-based approach just as audio practitioners have produced with surround sound. Live production uses the resources of the Control Room or OB Van mixing console and the audio mixer listens from the monitoring of the desk. See Chapter 4.

Live immersive audio production has been successful since the 2016 Olympics with NBC, as well as with NHRA using Dolby Atmos (see Chapter 10, Case studies). Production audio from the OB Van or Control Room is sent in discrete channels to a Network Operation Center (NOC), where it is integrated into other content and advertising and then the signals are sent for mastering by another black box for transmission encoding.

The implementation of future ready entertainment with integrated, immersive, interactive and intelligent features is well underway but broadcasters, equipment manufacturers and practitioners are still refining the tools and practices. The equipment manufacturers have fulfilled the basics with real-time mixing consoles and encoders/decoders have been implemented and new features roll out regularly, but still have work to do.

One of the problems is stimulating demand for next generation entertainment and working directly with the consumer market for advanced home systems. Content stimulates demand and the OTT broadcasters like Starz and Hulu have led with advance 7.1 and 7.1.4 sound designs on their episodic and drama programs.

Notes

1 Audio Definition Model (ADM) standardized by the ITU-R as BS. 2075-2.
2 High Efficiency Coding and Media Delivery in Heterogeneous Environments – Part 3: 3D Audio, ISO/IEC Standard ISO/IEC 23008-3:2015, 2015.
3 Silva, Robert, *User Guide Creating Multi-Dimensional Audio Assets: Overview of the DTS:X Surround Sound Format, Experience immersive surround sound with DTS:X*, www.digitalcommons.calpoly.edu/cgi/viewcontent.cgi?referer=&httpsredir=1&article=1049&context=laessp (accessed: January 13, 2021).

9 Convergence the Experiences

The consumer experience is available and embedded in future ready entertainment hardware and technologies along with the creative software and talents of the producers and programmers that flip the switches and pull the leavers that craft the magic. Virtual reality (VR), augmented reality (AR), mixed reality (MR), extended reality (ER) and 360 Video are proven experiences already in use for entertainment and education, but it still seems that large high resolution fixed screen monitors continue to be the picture and viewing experience of preference. The New Realities Experience is resulting in a complete convergence of different electronic delivery platforms that will include ultra high definition (UHD) pictures with interactive touch screens, motion control, immersive sound along with headphones, goggles, gadgets and more – true integration.

Not only is there a paradigm shift in entertainment technologies, but also production and storytelling methods. For example, with the old storytelling methods the sound designer would let you hear what they wanted you to hear, but with the new interactive experience the user controls what they want to hear.

The entertainment, education and information experience will accommodate different levels of connection, involvement and participation. As you progress from general background information to specific relevant information an individual consumer attention quotient changes. Background noise floods our brains with listening points – messages which are filtered and you react to. At the opposite extreme is education, where the consumer generally must focus on the content and media and is almost compelled to believe the content as factual, however part of the Entertainment Experience is also about wanting to believe.

The imperfections in the entertainment experience have been the interface between the content, experience and the consumer. Fortunately, the experience can be fooled. Consider that since the acoustic Gramophone was introduced to the public by Edison there has been endless curiosity at how believable it sounded only for consumers to be astonished at the next new entertainment and technical miracle. The next generation marvel is here, and it is a holistic approach that incorporates the concept of future ready entertainment and the principles of Next Generation Audio as proposed in 2016. Significantly, future ready entertainment incorporate features not described initially by NGA such as haptic stimulation, gesture and motion interfaces along with a commitment to develop features not even proposed yet.

Television is the conduit for information, education and entertainment, but consider that sound is a tricky subject because television can be viewed in the background with the sound off. Sports bars! What can I say? The acoustic environment is cluttered with background noise and 12 television sets in a bar would only contribute to the chaos; besides there are solutions to every sound situation. Why can't I use my personal device to listen privately to any one of 12 televisions or even two – one in each ear. Television entertainment has traditionally suffered from poor home sound reproduction, particularly with surround sound. Television audio has deteriorated considerably since the screen size has been stretched to the edges and television

DOI: 10.4324/9781003052876-9

manufacturers moved the speakers to the rear where there is no direct sound reaching the consumer.

There is the argument that some people don't care about the sound. I would maintain that no one cared about color television until they experienced it. The content side of the entertainment equation is demand driven while new devices are driven and sometimes miss-driven by manufacturer. I think a narrow-minded repudiation of past problems and not a forward glimpse at the new possibilities of future ready entertainment/Next Generation Audio (NGA) is a mistake. Things like headphones and headtracking are certainly changing the consumer side of the acceptance formula and clearly there will be greater future demand for the sound to fill a broader entertainment zone.

It is the responsibility of the content producers to offer a scalable entertainment experience and for the manufacturers to provide a wide range of user-friendly consumer devices and interfaces. The scalable experience will vary between users. For example, I would prefer an engulfing experience with my sports while you might listen to the commentator's audio only in one ear devices and pay for the coach's audio in the other ear. Both are immersive experiences.

Television sports is often a communal gathering for entertainment and socialization. It was always difficult for me to accept that televised sports is often a social event and the way television sports is produced the viewer never misses anything, often thanks to the sound. Imagine this scenario, you are enjoying your favorite team having a good time interacting with your friends when all of a sudden the roar of the crowd from the television captures everyone's attention for a series replays and commentary rehash – nothing missed.

With sports consumption the consumer decides what level of engagement and entertainment they want. The television experience is scalable while the theatre experience remains unique because you go to the cinema for the big bang, full sit-down, maximum attention experience – usually. Turn off your cell phones please. But consider that Virtual Reality is a complete immersion with the cocoon experience where your eyes, ears and brain are fully and electronically engaged – absolutely nothing missed, but no social interaction in the real world – yet. There is a vast selection of entertainment options evolving from fully immersed to full background and all points in between. For example, my Apple watch might alert me to a story I am following so I can adjust my personal listening device and focus my aural attention to the news, no need to watch talking heads and I can keep working on my computer.

The use of ear devices or headphones for sound augmentation for the hearing impaired and even for the listener that just wanted to cut out some of the background noise has been around for a while. There seems to be an abundance of home television type headphones, but there has not been much evolution to change the way these type of headset interacts with outside the headset environmental sound like a noise-canceling headset does. Speakers and acoustics will never be perfect, but audio augmentation over ear receptors could solve many problems – including speech clarity.

Several years ago, I was touring a manufacturer of hearing devices and presented the idea of the sound of a radio or television program transmitted directly to the consumer on a hearing device. No problem, but now make that hearing device capable of receiving natural sound from the soundspace immediately around the wearer – somewhat like the principles of a noise canceling headset, but electronically different. The earbuds or headphones can be adjusted through your smart device for the balance between the broadcast sound and the environment, plus some DSP trickery so the wearer could interact with other people and not have to take their ear device off.

The sound of television will benefit from the growing consumer acceptance of headphones and earbuds simply because hearing devices are a good way to get quality dimensional sound to the consumer.

The Convergence of Old Reality: Broadcasting and New Reality, MindCasting, Direct Sensory Shaping

The integration of traditional storytelling with pictures and sound, as we have done for centuries with next generation storytelling, will include multi-dimensional sensory stimulation, enhanced user interface and beyond. The entertainment industry is buzzing about the New Realities and audio will clearly benefit because this represents new opportunities for sound practitioners and studios.

The New Realities

VR, AR, MR and ER propelled in Headtracking and Headphones into the consumer entertainment zone and without a doubt VR audio production presents tremendous opportunities for the creative sensory/sound designer. Just as a director and producer use visual elements to tell the story, audio in a 3D world can be used to steer the attention of the consumer to various elements the sound producers want to draw attention to.

VR audio can be a combination of captured and created audio, however the audio track must be created from scratch and fill every sonic requirement for a believable immersive experience. The complexity of the 360 scene will dictate the extent of the sound design, however I will always recommend that you place an ambisonic microphone as close to the camera as possible to anchor your soundscape. Remember the 360 camera and microphone are stationary. A benefit for the sound designer is that immersive audio for VR and speaker-based reproduction is produced similarly using ambisonic production techniques which should represent a growth opportunity for the audio industry in both studio production and live capture.

All reality-based productions (VR, AR, MR, ER) offer tremendous potential for sports and entertainment in the form of an added reality adjunct to the main broadcast. For instance, when the broadcast production is scripted and the primary screen is center stage for the event then the content producers can add on enhancements like goggles, headphones, haptic motion enhancements and alternative cameras to the existing story and the story highlights can be directed from the main screen and commentary/hosts. Bobsledding is a sport that is presented from outside the sled and the program could give a rare glimpse from inside the sled. The commentators could tell the audience to put on their goggles and headphones for a ride down the track short and a 3D experience.

Many people associate augmented reality as a goggle experience and it can be, but after little thought I can see how it actually has interesting entertainment potential completely complementary to the live or screen experience. Just as second screens and event screens are powerful tools in the entertainment experience, the augmented audio experience is perfect for consumers, producers and marketers.

All the New Realities have different audio production requirements. AR is usually as simple as adding something to the reality base that you are already experiencing, as opposed to the cocoon-like VR experience. With augmented reality you generally think of images and graphics over reality, but audio over ear buds and a smartphone has the potential for augmented audio to span every aspect of entertainment because of this handheld portability.

Not only can augmented reality audio deliver better clarity and detail to the live sound experience, but it also delivers enhancements like commentary and translations as well as personal sound from athletes and coaches. Augmented audio reality is a beneficial sensory dimension for virtually any live production as a supplement to in-venue presentations. Augment audio can be creatively used in a variety of possibilities including live music, drama and theatre not only to supplement the PA audio but also customize the mix. Your audience is armed with a Wi-Fi enabled streaming device, their smartphone and earbuds. Augmented audio streams can be interactive, allowing the listeners to select alternative commentators, coaches, and athletes, inside the stadium or on the field, and have it delivered to their hand-held device.

360 Video, VR and ER are headphone base and are not interrupted by room acoustics or the human anatomy. We listen to stereo and binaural sound over ordinary headphones and earbuds, but binaural is different because it attempts to capture the direct sound, reflected sound from the physical perspective of the head and ears which factors in each person's indi-vidual anatomy and hearing characteristics.

Binaural audio is often thought of as sound that has been captured to create a 3D effect and there are countless convincing recordings of music, radio and televisions that have used the technology. Most of these sound productions used the traditional approach of either a manikin or real body with miniature microphones in the ear opening, however binaural rendering tech-nology allows for the creation of immersive sound experiences for the listener over headphones, which is it greatest potential. VR and augmented reality audio benefits from the use of ambi-sonics audio because of headtracking and high-quality sonic rendering of binaural sound.

Virtually all audio production software is capable of binaural reproduction that simulates the acoustic information of the direct sound, the environment and the filtering our body, ears and brain do. The biggest complaint about headphones is the in-head experience. What is missing when listening over headphones? Reflected sound and the anatomic filtering of the sound on the way to the brain. Our brain is sensitive to a less than realistic soundspace and natural binaural reverberation is important to the believability of the reproduction.

Immersive sound production techniques already use advance space simulation software which reproduces well over headphones. Real-time binaural synthesis using HRTF/ARTF (anatomy related transfer function) filtering and modeling continues to evolve where a listener will be able to audition a HRTF setting and pick the most realistic to them. Electronic and aural localization mimics psychoacoustic phenomenons such as reflective sound and HRTF calcu-lating desirable timing, amplitude and frequency characteristics of sound entering either ear.[1]

Positional tracking simply is the ability to freely look around a virtual or augmented world experiencing an immersive and interactive way to explore those virtual soundfields. Positional tracking registers the position of the head in space, recognizing movement front and back, up and down and to the sides, increasing the connection between the physical and virtual worlds. Head tracking (rotational tracking) indicates the rotation of the head transmitting represen-tative data about pitch, yaw and roll. Head tracking is certainly a potential paradigm shift for broadcasters while a mainstay in the realities world.

For 360 Video, and all the New Realities positional tracking with 3D audio locked to the image is essential to the illusion. Ambisonics, MPEG-H and Dolby Atmos all profess to pro-vide for accurate head tracking, but I believe ambisonics is the only technology with full and accurate positional head tracking.

Dolby keeps portions of the audio, such as the dialog, anchored to a specific location while you move your head. Clearly keeping the dialog fixed reduces confusion to the viewer/lis-tener. Dolby Atmos VR Transcoder transcodes a Dolby Atmos File (.atmos) to a Dolby .ec3 bitstream audio only .mp4 or B-Format, wav (FuMa and AmbiX) for mixing with video files.

Live and Streaming Sports: The Convergence of Old Reality, Broadcasting and New Realities, 360 Video and NGA Next Generation Audio

There is little wonder why a technology giant like Intel gravitated into the entertainment industry. Computer chips with extremely fast processing speed are a significant part of the future of entertainment. Intel has partnerships with NFL, NBA, MLB, PGA, NCAA and the Olympics to deliver spectacular Re-Play technology for broadcasting and have tested the water for live sports production. Intel made tremendous headway into sports with its Intel freeD Tech (Replay Technology) where 38 cameras mounted around the venue and stitching software that create a virtual camera that shows all angles. Sound? Typical replay sound – none.

Intel's True View uses Stereoscopic Camera pods with 12 cameras – six for the left eye and six for the right. Video 180 is a semi-circle viewing range rather than a full 360 degree – 180 creates the highest quality resolution of the field of play. Monoscopic 360 video is a single channel where Stereo 360 Video is two different channels which deliver more depth. Intel uses multiple cameras – pods – and can switch between the camera pods.

Intel's True View works on Samsung Gear VR headsets and lets you select from different camera positions.

Olympic Broadcast Services, a division of the International Olympic Committee tested VR production with Intel (a sponsor) at the 2018 Olympics with distribution platforms on NBC Sports VR and Eurosport using their respective Apps that work at home and on smartphones. Intell produced standalone VR productions from ceremonies, figure skating, speed skating, snowboard, half pipe, moguls and more using up to five camera arrays that have been previously described. While switching between the cameras brings a new dimension to sports viewership and certainly gives you an appreciation for behind the camera production, I do not think you could enjoy an entire sports production using goggles. I clearly see VR and AR as an add-on to the broadcast production.

I was the sound designer at figure skating and know that Intel Sports took the fixed audio from the host broadcaster which was either stereo or surround. Sadly, the audio production was nowhere near to what its potential was. Reality sound design is under development, however at the time of the publication of this book, the merging of broadcast sound and new reality sound has just begun. See the case study on figure skating in Chapter 10.

What is New Realities Sound Design? Capture, Create or Both?

Mixing mono, stereo and array microphones is a familiar and similar production workflow for senior level sound balancers and mixers. This same workflow would be used for an immersive sound production only adding a height layer making immersive and HOA production truly seamless for the mixer and audio producer. 360 audio and video production can easily be created in the post-production environment, however live reality production integrated with reality-based productions can extend the entertainment experience far beyond the consumer's expectations. As 360 video for sports evolves and matures, the sound reproduction will move quickly from a basic two channel production to truly immersive and interactive.

What are your sound design goals? What is your format? 360 Video, VR or AR. What are your video sources and synchronization essentials? Since the sound for all non-reality-based media can be captured or created from the ground up, the sound design is critical to the effectiveness of the illusion. For example, in real life when you turn your head the orientation of your ears alters the way things sound. The ability to recreate this level of interaction may or may not be essential to some VR, AR, MR or ER experience.

In sports, the PA and venue atmosphere are basically inseparable and a satisfactory balance can be achieve with good microphone placement off-axis from any speaker clusters. A sanitized atmosphere and a separate direct PA feed should be mixed together into the upper immersive speaker soundfield. Since ambisonic creation does not require ambisonics capture, the VR mix could ingest the mono and stereo microphones and generate a non-specific ambisonic atmosphere mix bed with head tracking capability. The ambisonic microphones at each camera position plus the non-specific ambisonic atmosphere mix provide several slightly different basic immersive soundfields and combined create an acoustic fingerprint of the entire venue along with a stable sonic foundation. Remember, certain reality experience can easily cause motion sickness.

Certainly the music in figure skating is the emotional and spiritual foundation of the athlete and performance. However the music is everywhere and often at significant levels resulting in PA music bleed into every single microphone. A good microphone design places microphones as close as possible to the desirable sounds and as off-axis as possible to the PA. The goal is to get enough separation between the PA and the direct sound to mix a dimensionally enhanced upmix of the music back into the general sound mix. Unfortunately all of the music is provided by the athlete and is usually a stereo mix which needs to be up-produced into an immersive soundfield. See case study on figure skating in Chapter 10.

The role of the sound producer at most sporting events is significantly different than the role of the VR sound producer. The VR sound producer spends time crafting out many immersive audio stems because the sound producer does not know what the consumer is looking at. The broadcast sound mixer also spends time creating immersive sound stems, but also injects the mixers their own personalization into a director-guided production and sound mix while with VR the consumer does the blending or mixing by head tracking and interaction.

Balancing sound element by the consumer is a possibility with Future Sound and Next Generation Audio (NGA) codex. Since ambisonic soundfields can be layered on each other by creating a separate soundfield for the sports-specific sound separate from the ambiance sound and separate from the music which gives many interactive options to the viewer/listener. Of course there is latency and there needs to be due diligence to make sure the sound and picture are in sync, which will be the responsibility of the audio production team. The VR sound team could create ambisonic stems that are actually mixed by the user in the home device to ensure sync.

Since the computer industry realized that VR production for gaming and sports was a great marketing tool for fast computers, Intel has committed to fast processing and storage and considering the fact that audio processing is so small compared to the processing required to produce the picture leaves plenty of time to sync the picture to the sound.

4D: Haptic

Haptic is the science and technology behind touch and tactile feedback and has a place in 360 Video, VR, AR, MR and ER sound production. I spent a decade mixing and producing sound for motorsports and when surround sound was implemented it was immediately obvious the benefits of the LFE to motorsports. Remember LFE means low frequency effect and not another form of bass management. Since low frequency rumble can be particularly annoying, effective use of bass frequencies needs to be carefully thought out.

First, haptic and using the LFE can often create similar sensations making them comparable, but the creation process is dissimilar. Consider this – Haptic should be cogitated as mechanical while LFE is strictly acoustical although often achieving similar impact. When 5.1 surround sound was introduced to motorsports it was obvious that routing a portion of a microphone signal away from the main mix and bandpassing a very low selection of frequencies from specific microphones and then gating the sound was a very powerful use of the LFE. Gating the sound was critical so that the sound would only open when a car was very close and not stay open, thus adding objectionable rumble to the overall mix. There is nothing more detrimental to a mix than excessive rumble that does not contribute to the audio production. In subsequent years a few audio practitioners tried using a dbx 120 Subharmonic synthesizer with some success, however at the time of writing I am not aware of anyone using this method.

The ability to feel the low end without an increase in volume from the LFE or subwoofer is extremely desirable and the foundation of the Guitammer company. Because the transducer is attached to an object like a chair or to the floor the effect is dependent on tactile feelings and motion and not auditory cues although the premise of the effect is converting electrical images – sound – into a mechanical vibration. Converting sound to low frequency vibration is the same principle that I applied to motorsports in the 1980s but the Guitammer "buttkicker" device does not reproduce sound. The transducer reproduces the frequencies between 5 Hz and 200 Hz with only vibrations and no sound.

Haptic is an area that was explored by several media production companies and found its way into the production of NHRA drag racing with Mark Luden and his Guitammer company and butt kicker product. Haptic worked particularly well with drag racing because of the intensity of the sound and the short duration of the effect. Since buttkicker is designed to use live the sound designer/mixer must limit the amount of low frequency information sent to the unit to minimize any unwanted and unnecessary vibrations. This can be accomplished with minimizing the input sources, having a person mix between the input sources or using a "gate" to automatically restrict the input sources.[2]

Motion Control and Gesture Control

The Enhanced Entertainment Experience is well underway with research into the human computer interface which has evolved from a typewriter-like keyboard to voice recognition. Gesture-based hand control appeared as early as 2005 with a Samsung phone that allowed the user to dial the phone by writing the numbers with hand gestures. Later that year Nintendo Wii became the consumer product that brought the concept of hands-free Motion Control to the public. Undoubtedly hands-free motion control and interaction proliferated in popularity because, since the iPhone 4, there has been an accelerometer in every iOS device that detects changes in the position of the device.

Gesture-based interaction such as page turning and zooming is common with touchscreen devices and completely possible with gesture recognition. Precision hand control can extend to XYZ axis mapping for object localization and can be extended to control mixing and panning functions in free space.[3]

Depth cameras for computer vision, body tracking and gesture recognition are only going to develop further with three-dimensional storytelling. Anatomical positioning will extend to future dimensional audio production where audio sources + metadata of user position is rendered binaurally to headphones.

360 Audio Production and Reproduction for Streaming on Non-Professional Devices

There is a blur between semi-professional equipment and a higher priced brand. Certainly you pay for durability with Pro Gear, but performance is questionable, making consumer software from Google, Facebook and Samsung viable, particularly because of their massive distribution platform. Google ARCore software and Apple Arkit is a set of tools that simplify building an AR experience, scene capture, motion tracking and scene processing. Apple's recognizes the spaces around you and allows you to place virtual objects.

Samsung (Milk) VR specifies Binaural Audio as 4 mono or stereo tracks and Milk VR constantly mixes only two of the tracks for headtracking. The audio engine (renderer) mixes between the tracks till the viewer reaches a quadrant fully. For example, if the viewer is looking between the quadrants, say at 45 degrees, the renderer will play out 50 percent of the first track and 50 percent of the next track till at 90 degrees 100 percent of the second track is played out. Quadraphonic uses 4 mono or stereo tracks, however Milk VR will play out an audio track for each of the 90 degree directions a user looks at. The renderer will fade between the tracks as the listener crosses into the next quadrant. The bottom line is that Samsung Binaural will constantly mix between two tracks at any given time while quadraphonic will only play out one track at a time.

Interactive Audio – EEVO is a 360 Video editing software that can make a video interactive. What EEVO calls "hotspots" is an area where the viewer can look at and make a sound element playout full volume, ducking the other audio material. Hotspots are drop and drag pointers and seem to be easily located in very specific locations. Once your sound source is identified you can spatialize the sound to the location in the scene.

For example, if you had a scene with a car, you might want to create a separate sound source for each wheel, and for the engine. As you move around the car, the placement of these sources solidifies the illusion of the VR space you have created. If this car was in an enclosed space, like a garage, you would also want to apply some kind of spacializer filter for each component to add a realistic sense of being in the garage.

Facebook's software "Two Big Ears" permits you to mix on a headset; Oculus audio and dearVR have proprietary rendering tools; and Jaunt is Cloud-based VR production suite and publishing tools. Whatever the tools used, consider integration across multiple platforms and manufacturers.[4]

Next Gen Entertainment: Beyond Viewing

I clearly see the integration of VR, AR, MR and ER into the storytelling aspect of a linear live sports and entertainment production. The profiles, personalities, trials and tribulations can be told effectively in a goggle and headphone experience. A production technique essential to immersive storytelling is to use 3D audio to guide our ears through localization information from all directions and distances. VR soundfields can be combined with specific sounds to steer the attention of the listener/viewer. Directing your viewer's attention is a powerful and nondestructive way to create an immersive and interactive experience.

There are similarities in the application of 3D audio and the sound design of immersive sound for VR/AR Headphone reproduction and speaker listening using ambisonic production practices. The ambisonic sound design uses natural capture, processing and mixing of the audio components to create an ambisonic sound foundation.

Production and post-production practices are in development but a flexible sound format is needed for the joint production of 4K, and VR and to me ambisonics-based audio production is applicable to immersive audio production for broadcast and broadband as well as all alternate reality-based audio productions. Additionally, ambisonics audio production works particularly well to produce Binaural Audio which is critical to the production of Virtual Reality Audio.

Future Proof: Feature Proof

There is no doubt that broadcasters will embrace a range of features to attract and keep future generations of consumers. Content is still king, but there has been a significant paradigm shift in the capture, production, mastering and distribution, not only across delivery platforms but also physical platforms.

Notes

1 Cheng, Corey I. and Wakefield, Gregory H., *Introduction to Head-Related Transfer Functions (HRTF'S) Representations of HRTF'S in time, frequency, and space*, University of Michigan, Department of Electrical Engineering and Computer Science Ann Arbor, Michigan, U.S.A, Systems and methods for HRTF personalization Patent No. US 10, 028, 070 B1 (45) Date of Patent: Jul. 17, 201.
2 Luden, Mark A., Interview with The Guitammer Company, Haptic.
3 Joseph J. LaViola Jr, Ernst Kruijff, Ryan P. McMahan, Doug A. Bowman, Ivan Poupyrev, *3D User Interfaces: Theory and Practice* (New York: Addison Wesley, 2nd Edition 2017).
4 Nonprofessional 360 audio www.production.arvr. www.google.com/arcore/developer www.apple. com/documentation/ www.arkit.eevo.com.

10 Case Studies

Sports

This chapter provides an in-depth look at how to produce immersive and interactive sound for more than 60 different sports. These case studies include immersive sound design philosophy, microphone selection, placement and mixing options for immersive sound plus the multiple derivative mixes that are required.

Case Study 10.1

Immersive Sound for Aquatics and Diving

Production Philosophy for Immersive Sound

Aquatics and diving coverage is usually presented from a variety of camera angles – with views looking up at the diver as well as a profile view and from above. The cameras emphasize the distance/expanse of the diver above the water. There is clear dimensional separation between key components of the sound origination points – specifically the platforms above and the water below.

Significant to the ability to create natural sounding dimensional soundscapes is the ability to place microphones close to the athletes and separate/isolate specific details in the cluttered soundfield from general ambiance and noise. Appreciably contributing to the illusion of height are those distinct sounds that are visually associated with different parts of the field of play – the diving platform, the springboard and the water.

The Aquatics – diving immersive sound design effectively uses the principles of Front Vertical Soundfield Enhancements (discussed in Chapter 4) which correlates and reinforces specific sounds in the front vertical plane – the top-to-bottom aspects of the sound mix as well as the front left-right aspects of the soundfield. Essentially detailed sound is dispersed into the entire anterior vertical plane of the viewer/listener.

From the side perspective, the athlete clearly appears as though they are above the heads of the spectators and specific sounds from the platform such as breathing and feet are not distracting. The elevation of the springboard is subjective and can be contoured with different types and location of microphones.

Microphone Placement

Immersive sound design for diving is built on microphone placement to capture an acoustically detailed soundspace by capturing and conveying specific sound details and their relationship in the sonic space. For example, there are two percussive aspects of the sound that will reverberate through the room. The springboard and the impact into the water are emphasized

DOI: 10.4324/9781003052876-10

Figure 10.1 Aquatics: Swimming Lanes

Figure 10.2 Microphones along the handrails at diving

at ear level with close microphones, however the delayed soundfield reverberates inside the venue and when captured and reproduced creates a stable immersive soundfield.

Sound Capture zones include the platform and board and water level splash, bubbles and venue acoustics. The platform and the springboard provide adequate space and profile to place stereo or pair of microphones with a low profile such as boundary, PCC or PZM type of microphones. The springboard is a source of very specific sounds with a wide dynamic range. A multi-capsule microphone such as the Audio Technica AE5100 dual capsule microphone captures a close spatial image. The splash microphones located symmetrically around the landing area fill the left, right, left surround and right surround speakers with direct, diffused and delayed splash sounds.

Mixing: Immersive, Surround and Stereo

Immersive sound emphasizes the low dramatic camera views simply by elevating elements of the sports sounds into the upper stratus

Dr. Barry Blesser wrote in his book *Spaces Speak* about the symbolic nature of acoustic space and that every physical space has an acoustic fingerprint. For example, the memory of the sound inside a cathedral, cave or swimming pool is vivid with expectations and by applying room acoustics to the direct sound a believable stable immersive soundfield is created.[1]

Hydrophones have been used effectively to capture underwater and can add another dimension to the immersive mix. When the athlete is completely underwater the mix goes underwater as well. Use a Spatial enhancer and create an immersive soundfield from the stereo hydrophones as well as create an LFE effect for the underwater soundfield.

Atmosphere – most aquatic venue designs have at least two or three tiers for seating, dimensional crowd layering is possible from microphone placement in front of the crowd and suspended from above the crowd. The Atmosphere perspective will not change during the production.

A dimensional ambiance can be derived from microphones in front of and above the audience plus a subtle blend of PA. Even though there are several different visual perspectives, the sound perspective will stay fixed.

Mixing with 3D mixing consoles

Real-time sound/audio sources can be mixed using typical touch-interactive channels and faders and routed to an immersive soundfield buss, and/or a surround soundfield buss and/or a stereo soundfield buss separately and/or simultaneously. Specific sound elements and microphones can be easily positioned in an immersive soundfield using XYZ panners similarly as the sound element can be easily positioned in surround soundfields with surround and stereo soundfields with panorama controls.

Mixing Immersive with Surround busses: Microphone Double Bussing/ Double Microphone Placement

> Ear Level Layer #1 – 5.1 Springboard microphone – L and R, front and back
> Ear Level Layer #2 – 5.1 Water splash and underwater microphones
> Ear Level Layer #3 – 5.1 Camera microphones
> Ear Level Layer #4 – 5.1 Atmosphere microphones – L, R, LS and RS
>
> Height Level Layer #1 – 5.1 Springboard microphones – L and R, front and back
> Height Level Layer #2 – 5.1 Atmosphere microphones – height left, height right,
> height left surround and height right surround

Height Level Layer #3 – 5.1 Atmosphere microphones – Side Height Left, Side Height Right,

Side Height Left and Side Height Right

Note: Ear Layers 1 and Height Level Layer #1 share microphones

Case Study 10.2

Immersive Sound for Archery

Production Philosophy for Immersive Sound

With any sport sound design a sense of aural space around the picture can be as simple as the addition of ambiance and atmosphere into the height speakers. Like in many case studies, archery sound does not have any visual support for sports sounds above the viewer/listener but the entertainment and production characteristics allow for colorful and interesting detailed sounds to enhance the production. For example, the swish of the arrow has been a dominant production sound for decades, with immersive sound elevating the swish further reinforces the illusion. Remember Dr. Blaurt wrote about high frequency sound above the listener and how this contributes to the impression of height (see Chapter 4). The height channels are used to fill out the immersive soundfield from deadspots, what the film industry calls room-tone and for emphasis and effect.

Microphone Placement and Mixing Techniques

The sound design for archery is built on an acoustically detailed soundspace with microphone placement specific to capture and convey motion. A line of stereo boundary microphones are placed between the archer and the target. The motion of the arrow from the archer to the target appears to go forward and can be positioned in the soundfield to move from back to front in the left and right speaker, by capturing and conveying specific sound details.

Figure 10.3 Boundary microphones in line with the path of the arrow in archery

To further emphasize the arrow and its relationship in the sonic space, elevation should be applied using one of two methods. Elevating the boundary microphone directly or by bandpassing the high frequencies and elevating those specific frequencies.

The immersive soundfield sequence begins the sonic space around the athletes preparing. This perspective should be a looser impression with full 360 reproduction. Next in the sequence is the motion of the arrow which is dimensionally over the present spacious soundfield. The soundfield abruptly changes to full forward vertical the instant the arrow hit the target. The impact of the target is emphasized in all forward speakers – left, right, left height and right height and is an example of using the height speakers for effect. Microphones are placed in the target as well as behind the target for a slight delay.

The spectators at archery are generally reserved during the period before the arrow release but are very lively after the arrow breaches the target. The Atmosphere swells into upper and lower level of the soundfield and slowly decays as the next arrow is readied.

Most archery venues are temporary designs with at least two or three tiers for seating. There is sufficient space for crowd microphones in front of spectator seating. Dimensional crowd layering is possible from microphone placement in front of the crowd and suspended from above the crowd. The atmosphere perspective will not change during the production.

The PA is used to energize the crowd and there is an increase in the use of the PA for play-by-play in the venue. There is an air horn that signals play – no need to place a microphone for this sound.

Mixing with 3D Mixing Consoles

Specific sound elements and microphones can be easily positioned in an immersive soundfield using XYZ panners similarly as the sound element can be easily positioned in surround soundfields with surround and stereo soundfields with panorama controls.

Mixing Immersive with Surround Busses: Microphone Double Bussing/ Double Microphone Placement

Ear Level Layer #1 – 5.1 L and R, front and back
Ear Level Layer #2 – 5.1
Ear Level Layer #3 – 5.1 Camera microphones
Ear Level Layer #4 – 5.1 Atmosphere microphones – L, R, LS and RS

Height Level Layer #1 – 5.1 Atmosphere #1 microphones – L and R, front and back
Height Level Layer #2 – 5.1 Atmosphere #2 microphones – height left, height right,
            height left surround and height right surround
Height Level Layer #3 – 5.1 Atmosphere #3 microphones – side height left, side
            height right,
            Side height left and side height right

Note: The ear layers and the height level layers do not share microphones.

Case Study 10.3

Immersive Sound for Athletics: Shotput, Hammer, Javelin, Long Jump, High Jump and Pole Vaults

Introduction to the Broadcast Production of Athletics

Athletics is a multi-discipline group of horizontal – jumps, vertical and hi-jump, throws and running events with men and women competing in all disciplines.

During competition there are simultaneous events over the entire stadium/venue. This presents unique challenges and opportunities for audio capture and production. Each discipline and event support distinctly separate sound zones with sports specific and localized atmosphere sound that can be easily mixed and sculpted into an effective dimensional soundscape.

Production Philosophy for Immersive Sound

Athletics is a group of sports each with an opportunity for advance and interesting immersive sound design because of the ability to place microphones close to the athletes and separate specific details in the sound design from general ambiance and noise.

Distinct and separate sounds can be captured from each athletic zone and with spatial placement effectively implement the principles of front vertical soundfield enhancements (discussed in Chapter 2) which correlates and reinforces the front, top to bottom aspects of the sound mix as well as the front left to right aspects of the soundfield.

For example, from the side perspective of pole vaults, the athlete begins running at ground level and then vaults up to the top edges of the screen and lands on the raised mats. The mix begins with a dominant ear-level mix, then adds the height-level embellishments and ends

Figure 10.4 Microphones on short stands on the ground are physically spaced and can be spatially oriented to give a spread-out horizontal dimension to the sound and picture

with the landing at ear level full filling the front vertical soundspace. Pole vaults not only have a pronounced vertical separation but also horizontal motion between the run-up, vault and landing.

The athlete sounds from the hammer and shotput throws are captured in a relatively small space with a rope safety net around the action. The initial view of the athlete is head to toe and the athlete can be animated and vocal and moving at pace during the preparation. The athlete rotates 360 degrees with the hammer and releases the hammer through the opening of the net into the open field. Microphones are placed above the athlete and at foot level to create a compact 8.0 immersive sound zone around the athlete which is blended with a spacious 8.0 landing zone perspective.

Javelin and long jump benefit from significant horizontal movement but are dependent on atmospheric and ambient capture and enhancement for stable immersive soundfield.

High jumps has some vertical separation between the ground, landing mat and bar that can be spatially processed for separation and not localization.

There are a couple of different schools of thought for track, including some sampling supplementation. Track can benefit from the dynamic expansion of the starting zone and the full stadium atmosphere. The starting block is covered with up to two handheld cameras with stereo microphones which deliver a very close intimate perspective of the athlete and preparation.

There are several different local competitions and visual perspectives and each zone may have some sonic differences; however, the objective is to have some consistency between zones. Naturally each zone will have different audience reaction times as the sports competition progresses but each zone is off-axis of every other zone allowing the on-camera zone's atmosphere and ambiance to punch out and fill the mix.

A dimensional ambiance can be derived from microphones in front of and above the audience zone, however select microphone positions between PA clusters which are usually suspended. Unfortunately, there usually is a substantially active PA announcer and loud PA system. The sound perspective of each zone will stay fixed during the entire event.

Microphones and Mixing

The microphones in the throwing cage are placed around the athlete above and at ground level and by their physical positioning and arrangement they effectively create an immersive soundfield giving a height/sound aspect to the picture.

For the pole vault events miniature microphones are placed on the vertical bar suspension and on each corner of the landing mat. The four corners of the mat can be routed to the left, right, left height and right height channel for a full-frontal experience from the landing of the athlete. The running zone is covered by stereo shotguns on microphone stands, however they are spaced fairly wide because of the large area for the vaulter to prepare.

There are track events that generate some sound like the hurdles and water hazard but generally there is very little sound associated with running on a hard track surface. Typically, microphones are placed near the hurdles and hazard as well as the bell at the finish line.

Competition Zones and Atmosphere Production

Athletics is unique because of the large field of play footprint with up to five separate disciplines active at a time with simultaneous competition and coverage. Each competition is its own production zone with specific apparatus and localized atmosphere sound. A dimensional crowd is created from spatial separation, layering and purposeful placement of microphones.

Figure 10.5 Microphones clipped on the corner of all landing mats

---

Production Note

The Athletics Federation will want to approve the microphone placement and rigging and ultimately has the final say on where a microphone is placed and all issues and decisions on the field of play. Apparatus microphones and cables must be constantly scrutinized for safety and appearance.

There is usually a space between the field of play and the seating areas which simplifies crowd microphone placement. Crowd microphones should be placed on stands in this space to capture a present on-axis sound and along with microphones suspended over the crowd. A dimensional crowd layering is possible simply from microphone placement.

However, a large, open multi-sport production suffers from ambient noise pollution.

The ambient sound includes sports, crowd, PA and the reflections of everything inside the hard surfaces of the venues.

Mixing with 3D Mixing Consoles

Specific sound elements and microphones can be easily positioned in an immersive soundfield using XYZ panners similarly as the sound element can be easily positioned in surround soundfields with surround and stereo soundfields with panorama controls.

Mixing Immersive with Surround Busses: Microphone Double Bussing/Double Microphone Placement

Ear Level Layer #1 – 5.1 Apparatus microphones –L and R, front and back
Ear Level Layer #2 – 5.1 Floor microphones
Ear Level Layer #3 – 5.1 Camera microphones
Ear Level Layer #4 – 5.1 Atmosphere microphones – L, R, LS and RS.

Height Level Layer #1 – 5.1 Apparatus microphones – L and R, front and back
Height Level Layer #2 – 5.1 Atmosphere microphones – height left, height right, height left surround and height right surround.

Height Level Layer #3 – 5.1 Atmosphere microphones – side height left, side height right, side height left and side height right.

Note: Ear Layers 1 and Height Level Layer #1 share microphones.

Summary

The height aspect of the sound design will stay constant even with dimensional changes in the mix accounting for left–right and front–back camera perspective and orientation.

An aspirational goal is to create a holistic soundscape when cutting between five different events.

Case Study 10.4

Immersive Sound for Baseball

Production Philosophy for Immersive Sound

With any sport sound scheme, minimum immersive sound design can be as simple as the addition of ambiance and atmosphere into the height speakers which usually creates a sense of aural space for the picture. Like in many case studies, baseball sound does not have any visual support for sports sounds above the viewer/listener, but the entertainment and production characteristics of baseball allow for colorful and interesting support sounds to enhance the production. This sound design for baseball uses immersive principles which emphasizes and reinforces the height aspects of the mix that may be decoupled from the game aspects of the mix.

When you sit behind the home plate the "sound of baseball" is all around you. The fans and feel of a venue, the crack of the bat and umpire from the field. The venders hawking beer, peanuts and ice cream and the sound of the mighty organ leading the crowd in the familiar songs of baseball.

Cross-over appeal – the sound of the Hammond Organ is the signature sound of baseball. Get a direct feed, stereo if possible, and process the organ feed to immersive – Note you will have to delay the soundfield to match the acoustic soundfield of the organ playing through the venue PA system. Baseball stadiums usually have a significant amount of steep seating around the field of play which visually support the use of ancillary concession sounds, music and team-specific vocabulary positioned above the listener.

Microphone Placement

Immersive sound design for baseball is built on microphone placement to capture an acoustically detailed soundspace by capturing and conveying specific sound details and their relationship in the sonic space. For example, when you visit a baseball stadium the field of play is toward the inside of the venue while concessions and toilets are under the stands.

Most stadiums have at least two or three tiers for seating; dimensional crowd layering is possible simply from microphone placement.

Mixing – Immersive, Surround and Stereo

Atmosphere – most baseball venues designs have at least two or three tiers for seating; dimensional crowd layering is possible from microphone placement in front of the crowd and suspended from above the crowd. The atmosphere perspective will not change during the production. A dimensional ambiance can be derived from microphones in front of and above the audience plus a subtle blend of PA.

Mixing with 3D Mixing Consoles

Specific sound elements and microphones can be easily positioned in an immersive soundfield using XYZ panners similarly as the sound element can be easily positioned in surround soundfields with surround and stereo soundfields with panorama controls.

Mixing Immersive with Surround Busses: Microphone Double Bussing/ Double Microphone Placement

Ear Level Layer #1 – 5.1 Home plate microphone – L and R, front and back
Ear Level Layer #2 – 5.1 Camera microphones
Ear Level Layer #3 – 5.1 Atmosphere microphones – L, R, LS and RS.

Height Level Layer #1 – 5.1 Atmosphere #1 microphones – L and R, front and back
Height Level Layer #2 – 5.1 Atmosphere #2 microphones – height left, height right,
        height left surround and height right surround.
Height Level Layer #3 – 5.1 Atmosphere #3 microphones – side height left, side
        height right,
        side height left and side height right.

Note: The ear layers and the height level layers do not share microphones.

Case Study 10.5

Immersive Sound for Basketball

Production Philosophy for Immersive Sound

Clearly all microphone plans are dependent on the sport and venue but they are further confounded by what is the listener expecting to hear. With basketball the broadcast audience expects to hear the swish of the net in basketball even though that sound is rarely heard beyond sports broadcast or in film. The swish of the net is "Made-for-TV sound," satisfying the most basic sound expectation of the sport.

With any sport scheme minimum immersive sound design can be as simple as the addition of ambiance and atmosphere into the height speakers. This usually will create a sense of aural space for the picture, but advance and interesting immersive sound design requires understanding the point of view of the picture coverage to determine the POV of the listener.

This immersive sound design uses the principles of **F**ront **V**ertical **S**oundfield **E**nhancements (discussed in Chapter 4), which emphasizes and reinforces the front – top to bottom aspects of the mix.

Define your sound design by looking at the pictures. Basketball uses three main cameras. One game camera and 2× handheld – under the basket cameras. The game camera is behind the crowd usually capturing a shot of the full court from side to side. The pictures usually have a vertical aspect. The POV (point-of-view) of the play-by-play camera is often wide which can cover the action of multiple players including the basketball stanchions, backboard and net.

The under-the-basket handheld cameras POV is close-ups and often a head-to-toe shot of the athlete and referee on the court and under the basket area. Frontal vertical soundfield enhancements focus the attention of the listeners forward and uses the height element to draw the listeners' attention beyond the normal left and right peripherally. When you switch to the

– Front Vertical Zone =
Left, Left Height, Right Height, Right Speaker.

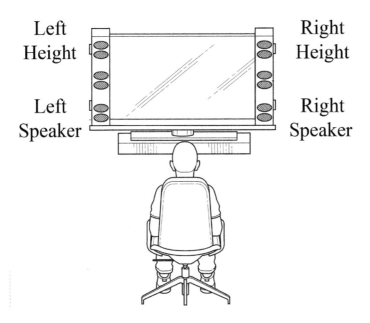

Figure 10.6 Front vertical zone = left, left height, right and right height speakers

goal camera, the net microphones are positioned into the height speakers and the close floor microphones appear slightly below ear level because the sounds are heavy in low frequencies.

The goal cameras presents an ideal head-to-toe picture and perspective for enhancing the 16×9 picture with top-to-bottom sports specific sounds in the front vertical soundfield/zone. The height aspect gives the sound designer the ability to place the sound within the four corners of the picture directly in front of the listener. Frontal zone sound image placement creates a correlation between the boundaries of the picture enhancing real and virtual phantom imaging.

Leaving the sound of the ball slap at ear level orients the slap of the ball at the bottom of the picture while adding elevation to the net microphones and keeping the swish at the top of the picture, helps the mind perceive an obvious vertical separation.

Additionally, medical research indicates that it is easier for the ears to hear sounds from in front and above which lends support to the use of frontal vertical soundfield enhancements principles in immersive sound design.

This basketball sound design uses microphones near the net, on the stanchions and fixed to the wooden floor to capture the FOP sounds. Basketball is a challenge to balance the sports sound against the ambiance and atmosphere with potentially excessive PA.

A spaced pair of miniature microphones are used to capture the net and under the basket sounds. The microphones are so close to the basket and athletes giving a distinct left and right sound image of all the action under the basket along with a dimensional net swish.

During reproduction the microphones should be elevated at least 30 degrees above the listener and panned to the left – 10 o'clock and to the right 2 o'clock using XYZ panners. With XYZ panners precise localization and proper sonic distribution between the frontal zone speakers can be easily achieved.

Figure 10.7 This picture framing lends itself to separating the crowd atmosphere above and the court bounces and squeaks in the lower section

Figure 10.8 A spaced pair of miniature microphones are used to capture the net and under the basket sounds

A similar effect can be accomplished by using multiple 5.1 busses to build an immersive sound bed. This effect can be accomplished using four 5.1 surround sound busses. The net microphone can be routed to an ear level 5.1 buss and a height level 5.1 (4) channel buss. The volume ratio between the height net microphone channel and the ear-level net microphone channel can be adjusted between the busses to find an appropriate level for the production. These ear level and height level effects busses can be further summed with additional ear level and height level 5.1 surround busses before being outputted for distribution and coding.

Mono shotgun microphones on each side of the stanchions have good reach into the midsection of the court. These microphones are positioned ear-level to the left and right of the listener using XYZ panners or assigned to the ear-level 5.1 surround buss. Additionally, the handheld cameras under the basket should be fitted with stereo shotgun microphones and imaged forward – panned left/right into the ear-level 5.1 surround buss.

Floor microphone can include boundary and/or contact microphones to anchor the low frequency sound ear-level and lower using some band-passing to the subwoofer.

Contact microphones have been discussed in Chapter 3. The contact microphone has proven itself to be effective in separating acoustic and resonant sounds common in basketball. Probably the most objectionable sound in television sports audio is the common basketball sneaker squeaks which are difficult to suppress with convention acoustic microphones.

Combinations of contact microphones have been tested using as few as four to as many as 30 during the 2020 NBA season. Consider that the ball sound resonates 360 from where the balls have contact with the wooden floor and the intensity of the sound will vary with the distance of the microphone from balls contact point. The difference between 4 and 30 contact microphones is the consistency of the ball sounds as the athlete and ball move back and forth on the floor. Sporting arenas and venues have been designed for multiuse and it is common for a basketball floor to be installed the day of the event making it possible for contact microphones to be mounted to the floor from underneath.

Talk to your producer and approach venue management about a plan to install contact microphones when the floor is installed and de-rigged. The NBA has experienced a different level of sound quality during the 2020 basketball season and along with the venue's management there should be more cooperation with the broadcasters to achieve a higher quality of sound.

There has been a growing trend to put microphones on the players which heightens the action and interaction between the players. Microphones on basketball players is problematic because specific language cannot be broadcast on certain television in the US plus in many parts of the world. The dialog is often recorded and/or cleaned up and replayed to avoid any issues. MPEG-H has interactive channels where the dialog could be put on its own interactive channel where any censorship would not be necessary.

At most indoor basketball venues natural atmosphere capture is difficult because of the typically excessive use and volume levels of the PA system. Microphones can be placed in front of the crowds on the stanchions *Photo*. This close placement of microphones to the crowd along with their slight off-axis orientation to the PA system should improve your crowd capture balance over the PA volume level. Clean, present atmosphere sound in the upper front sound zone contributes to the illusion of spaciousness around and above the listener/viewer. Another spaced stereo pair or stereo shotgun microphone should be mounted to capture more room, resulting in more ambient PA. These pairs of microphones and stereo microphones can be oriented either left–right, front–back or bottom to top easily giving dimension and cohesion to a 360-degree mix.

Sound Format Requirements: Derivative Mixes and Splits

If you are mixing immersive sound for an audio codex such as MPEG-H or Dolby ATMOS then all mixes from mono to immersive will be rendered appropriately. If you are mixing for discrete outputs many mixing console manufacturing include surround to stereo summing within the console. Note: these parameters are usually adjustable.

Combining the upper 5.1 layer and the ear-level 5.1 layer would be most effective by building separating 5.1 layers for specific effects, such as the net microphones to have level controls of the microphones in the ear and upper level.

Many broadcast level mixing consoles allow summing of multiple 5.1 layers over each other to create a finished composite of the 5.1 layers. For example: 5.1 Layer A and 5.1 Layer B and 5.1 Layer C can be combined to equal a composite 5.1 Layer Z.

Summary

Typical basketball coverage has three different picture compositions that are constantly switched between game camera and two basket cameras. Changing the sound perspective is not tedious or fatiguing to the viewer because the picture changes are accentuated by the height sound and variation in immersion experience.

The dominant immersive sound feature is the height/dimensional contrast between the net and floor microphones. This sound design lets the net sound punch out into the upper front stratus of the soundfield, unmistakably above the viewer and above the athlete in the picture action. The sports-specific net sound along with filtered PA and frequency-specific atmosphere in the upper front sound zone delivers dimensional and realistic sound to the listener/viewer.

Case Study 10.6

Immersive Sound for Boxing

Production Philosophy for Immersive Sound

Boxing is a sport that has been presented in full surround with sound behind the listener/viewer since the 2012 Olympics and this athlete point of view with fully surrounded sound design is still good practice for the ear-level soundfield in the immersive mix.

Boxing has a relatively small athlete field of play and the sports action is confined and relatively easy to capture. Generally boxing and table tennis have no "sports-relevant" sound in the vertical axis (above the listener) only ambiance and atmosphere, however both sports have opportunities for interesting immersive sound creation from microphone placement, capture and spatialization practices.

For example, the athlete, punch and referee sounds can appear to be fully frontal (top to bottom) from a listener seated at ring-side. This effect is present from either of four sides of the ring which will match any camera view and reinforce the use of frontal vertical – top to bottom sound design.

The foot sounds in boxing can be bandpassed, separated and mixed to enhance the spaciousness of the picture.

Over the boxing field of play there are four mono microphones hanging several meters above the ring's floor, pointed straight down at the ring for full coverage. During competition there are two handheld cameras in the neutral corners which capture the present sounds of

Figure 10.9 Boxing overhead microphones

the punches and the snorts and vocalization of the athlete. Additionally, there are microphones in the ropes, on the referee and under the mat.

The boxing ring is the platform for the capture of a dimensionally diverse set of sounds. Microphones on the net deliver a close perspective to the athlete plus microphones attached to and under the table contribute to a wide range of sound that is isolated and can be used for creating a dimensional soundfield.

The coaches in both sports are very animated during the competition and are often heard while the coaches appear on-camera and off-camera in the background. This is a sport where the coaches sounds are interesting to the coverage and relative to the competition.

Coverage of the boxing timeout is usually private time between the coach and athlete, however capture for broadcast only is accomplished with close stereo microphones on fishpoles or with miniature microphones in the ropes. With table tennis the athlete comes to the coach's area in the corner of their respective half of the court and microphones are placed in the "look of the games" barriers that separate the areas.

Spatial expansion (see plug-ins) of the stereo and mono microphones for dimensional soundfield enhancements along with the addition of ambiance and atmosphere into the height speakers creates an intimate sense of aural space for the picture. With this style of sound design, algorithmic spatial expansion fills any dead spots in the immersive soundfield and is not used for localization.

The front vertical soundfield – sound from the athlete point of view with the athlete and competition full in front is possible because there is clear separation between seating and the field of play in both sports.

There is usually limited ring-side seating at major boxing events, however there is usually a larger audience and more seating at table tennis. A combination of microphones in front of the seating stands as well as over and even behind the audience for some off presence atmosphere contributes to the immersive sound experience.

Overhead atmospheric enhancements do not require any localization and can be accomplished on most 5.1 mixing desks. Note: 3D panning is helpful for precise localization, but bottom line – convincing immersive sound for most sports does not require precise localization.

A focused field of play with a clear separation from the audience ambiance creates the complete "man-in-the-stands" soundfield with enhanced ambiance and a focused sports action sound.

Immersive sound does not need to be over the top to be effective. Sonic frequency expansion and band splitting is possible by routing microphones and combinations of microphones to separate surround or immersive group busses. Remember the Blauert effect of high and low frequencies.

Down Producing

There are differences in Codex rendering and multichannel summation for downmixing and down producing with different qualitative results.

Case Study 10.7

Immersive Sound for Cycling: Road Cycling, Velodrome, Mountain Biking, BMX

Immersive Sound Production Philosophy – Front Vertical Sound Field Enhancements

Immersive sound production for each cycling sport is different because each venue and field of play is unique. Road cycling is outdoor in a variety of different scenery, while mountain biking is outdoor in a variety of rugged terrain setting, BMX is out door with a distinct setting at the start through a dirt course surrounded by seating. Finally, velodrome cycling is surrounded by seating on an angled wooden track.

With any sport sound scheme minimum immersive can be as simple as the addition of ambiance and atmosphere into the height speakers but each cycling sport has a different height aspect and sound design.

Microphone Placement and Mixing Techniques

BMX Cycling

BMX has a unique starting zone that lends itself to a creative although upside-down side design. A row of cyclists launch past a metal start restrainer down a steep 20-meter elevated ramp. From the cyclist point of view the crowd is below him and the start camera sequence clearly emphasizes the elevation difference. There is a starter giving commands over a low-fi horn-type industrial speaker. The cyclist are often animated and play to the camera.

The start zone sound is captured from microphones at the feet of the cyclist, from a stereo microphone on the handheld camera and microphones on the structure above the cyclist – clear physical separation between the sound elements and the cyclist and equipment.

The ramp is designated a separate zone because of the unique sound of the cyclist moving from top to bottom. While on the course the immersive sound changes to typical person-in-the-stands with the course sounds dominant at ear level and the fan and PA banter in the upper zone. Each zone creates a sense of aural space for the picture.

The stadium designs at most BMX events are temporary and have at least two or three tiers for seating, dimensional crowd layering is possible simply from microphone placement. There is sufficient room to place microphones stands in front of the spectators off axis of the PA clusters – the PA at BMX is usually as loud as possible. Even though the PA is usually excessive, the upper layer in cycling benefits from strong PA sounds positioned above the listener because of the festive atmosphere.

Wireless microphones on competitors has been discussed with approval from the BMX Federation as early as the 2008 Summer Games; however it is challenging to wire each competitor.

Road Cycling

Most cycling sounds, except for BMX, do not have any visual support for sports sounds above the viewer/listener. Much of the coverage of road cycling comes from up to five broadcast chase vehicles with cameras and microphones that provide no dimensionality to the sound even though there is a significant emphasis on motion and a resulting Doppler shift.

Even though the Doppler effect contributes to the sense of speed from stationary cameras and microphones the abrupt sound of microphones moving past spectators can be objectionable and should be softened by trailing the sound into the rear speaker.

The most effective and successful sound design for road cycling includes a consistent and heavy use of samplers with spatializing software for cycle and crowd effects.

Ideally one sampler would focus on the changing scenery and spectators while a separate sampler would foley various athlete and cycle sounds.

Advance audio production can adjust the spatial properties of sound elements, but also change the tone and sonic characteristics of spaces.

Mountain Bike Cycling

Mountain biking is through a variety of terrain from an apparent remote location with no consistent background. This changing landscape lends itself to remote zones although between four and six microphones in a remote zone would be expensive to implement. Mono spaced pair of shotgun microphones to expand the capture zone with fewer microphones.

Birdie microphones – for example, the height channels are used to fill out the immersive soundfield from deadspots, what the film industry calls room-tone and for emphasis and effect.

Spectators around the course – lip sync may be an issue.

Velodrome Cycling

The sound design for velodrome cycling is built on an acoustically detailed soundspace with sounds definitely derived from the wooden track. Contact and boundary microphones as well as operators can track and capture the local sounds. Sound supplementation is recommended to add interest to the soundtrack.

Mixing with 3D Mixing Consoles

Specific sound elements and microphones can be easily positioned in an immersive soundfield using XYZ panners similarly as the sound element can be easily positioned in surround soundfields with surround and stereo soundfields with panorama controls.

Figure 10.10 Microphone suspended over velodrome

Mixing Immersive with Surround Busses: Microphone Double Bussing/ Double Microphone Placement

Ear Level Layer #1 – 5.1 Microphone – L and R, front and back
Ear Level Layer #1 – 5.1 Sample channels
Ear Level Layer #3 – 5.1 Camera microphones
Ear Level Layer #4 – 5.1 Atmosphere microphones – L, R, LS and RS.

Height Level Layer #1 – 5.1 Atmosphere #1 microphones – L and R, front and back
Height Level Layer #2 – 5.1 Atmosphere #2 microphones – height left, height right, height left surround and height right surround.
Height Level Layer #3 – 5.1 Atmosphere #3 microphones – side height left, side height right,
side height left and side height right.
Height Level Layer #1 – 5.1 Sample channels

Note: The ear layers and the height level layers do not share microphones.

Case Study 10.8

Immersive Sound for Formula E: VR and 4th Order Ambisonics Sound

The most compelling and entertaining VR sports demo was from Dr. Deep Sen and his team from San Diego. Pure ambisonic capture is tricky because the math predicts the high resolution in the audio capture, however due to background noise and often no clear natural separation of the sound the results can be a beautiful dimensional capture without much distinct resolution of its individual elements. Dr. Sen and I often discussed which sports might be possible with only a pure ambisonic capture without spot microphones and because Formula E is so unique because of the lack of certain car sounds there is a compelling soundfield within a few feet of the car.

Formula E is an electronic version of an open-wheeled racecar. Without the loud engine noise masking any other close sounds, there is an abundance of sound elements that are easily localized with a look. The short VR production is a perfect example of integrating VR into a broadcast production. The audience can put on their goggles and be there in the pits with the crew – then take off their goggles and back to racing to the big screen.

The POV of this demo is from one of the pit crew. The E-Car appears in the left periphery and stops directly in front of you. The sound of the pit stop is compelling and localized because it is not being masked by engine noise. You see and hear the tire jacks and changers to the left and right of you. You can hear the crowd over your shoulder and by turning your head it rotates the picture and the sound to full-on crowd with full presence and full frequency sound. Next your attention is directed to a helicopter overhead and then back to the car as it squeals out of the pits.

This demo works well because picture and soundfield are relatively compact and the audio is very dynamic and no single element overwhelms other distinct sounds.[2]

Case Study 10.9

Immersive Sound for Equestrian Arena, Dressage and Jumping

Production Philosophy for Immersive Sound

The two different equestrian arena events provide distinctly different immersive soundfield experiences. While the jumping events make available ample space for dimensional microphone placement, dressage will depend on ambiance, atmosphere and bandpassing the horse and athlete sounds to spatially separate the distinctly different frequencies in the immersive soundfield.

Jumping does have some visual support for sports sounds above the viewer/listener as each obstacle requires the horse and rider to be airborne. For example, during the leaps the rider and the horse are often vocal with distinct mouth and nostril sounds. These sounds are effective at dimensionalizing the picture when elevated in the mix.

Microphone Placement and Mixing Techniques

The sound design for the jumping events are built on the close proximity of the microphone to the rider and horse. Microphones should be placed on each side of the obstacle providing a spaced pair and stereo coverage of the capture zone. Additionally, microphones or transducers can be placed at ground level to capture low frequency sounds from the horse trotting, adding a sonic depth to the mix and helping emphasize the illusion of immersion and separation.

No additional approach and leave microphones are used because of the closeness of the obstacles and the difficulty of daily rigging and cable reset.

At dressage it is desirable to place microphones as absolutely close as possible to the field of play. Boundary microphones offer close proximity with a low visual profile and can be placed around the arena at ground level, hopefully between PA speakers. Microphones at ground level should give emphasis to the low frequency sounds from the impact of the horses' hoofs against an earth floor.

Stereo boundary microphones or pairs of mono microphones should be triple bussed – assigned to three different surround sound or immersive audio busses where one would receive a high pass filter, one would receive a low pass filter and one would not be equalized. The unequalized buss would be dominant in the front left and right speakers while the high channels

Figure 10.11 Microphone on stands around equestrian arena

would be positioned above the listener and the low channel below the listener and some to the LFE channel.

At many equestrian venues, stand seating commonly encircles the field of play. The audience is usually polite and quiet during the competition and jubilant at the end of a performance. Most equestrian venues are temporary designs with at least two or three tiers for seating and there should be sufficient space for crowd microphones in front of spectator seating and hanging from above. Dimensional crowd layering is possible from microphone placement in front of the crowd and suspended from above the crowd. The atmosphere perspective will not change during the production.

Be mindful of the positioning around PA speakers and that the PA is not used in the venue to announce the competition.

The immersive soundfield sequence emphasizes the sonic space around the athletes and this perspective should be a looser impression with full 360 reproduction emphasized in all speakers – left, right, left height and right height.

Mixing with 3D Mixing Consoles

Specific sound elements and microphones can be easily positioned in an immersive soundfield using XYZ panners similarly as the sound element can be easily positioned in surround soundfields with surround and stereo soundfields with panorama controls.

Mixing Immersive with Surround Busses: Microphone Double Bussing/ Double Microphone Placement

Ear Level Layer #1 – 5.1 Microphone – L and R, front and back
Ear Level Layer #2 – 5.1 Camera microphones
Ear Level Layer #3 – 5.1 Atmosphere microphones –L, R, LS and RS.

Height Level Layer #1 – 5.1 Atmosphere #1 microphones – L and R, front and back
Height Level Layer #2 – 5.1 Atmosphere #2 microphones – height left, height right,
        height left surround and height right surround.
Height Level Layer #3 – 5.1 Atmosphere #3 microphones – side height left, side
        height right, side height left and side height right.

Note: The ear layers and the height level layers do not share microphones.

Case Study 10.10

Immersive Sound for Equestrian Cross Country and Golf

Production Philosophy for Immersive Sound

Immersive sound production for golf and cross-country equestrian are very similar because the sound designer uses the consistent ambiance of nature blended with small crowds around the course to create a coherent immersive soundfield. With any sport sound, design a sense of aural space around the picture which can be as simple as the addition of ambiance and atmosphere into the height speakers.

The sound of cross-country equestrian does not have any visual support for sports sounds above the viewer/listener and even though the golf ball is often visually airborne the capture and reproduction of airborne sounds is virtually impossible.

The height channels are used to fill out the immersive soundfield, what the film industry calls room-tone – environmental ambiance free from dead-spots.

Microphone Placement and Mixing Techniques

The sound design for the cross-country equestrian events are built on an acoustically detailed soundspace with microphone placement specific to capture and convey motion – for example, hear and see the horses coming and going.

Virtually every camera picture must become an ambient sound zone of background fill, small crowds and the athlete moving from zone to zone plus there are at least two or three generic "birdie nature" zones without spectators. The Birdie microphone is the non-specific nature ambiance that is used to fill the gap between cameras or scenes.

With 18 holes in Golf and often more than 20 obstacles in cross-country equestrian both events are equipment-intensive events – microphones, cable and mixer capacity. Both events have sub-mixers generating stereo mixes of the various action zones, while the "show" mixer composites the stereo action sound with the immersive soundfield.

Golf usually always uses two microphones to cover each tee. This pairing acts as a very far spaced stereo pair and gives space and separation to the picture. Cameras on the fairways will usually use mono shotgun microphones to get further reach toward the athlete while greens will have multiple stationary microphones and possibly an operated and a microphone on the handheld cameras. All this stereo sound gives a strong forward image around the picture to be married to a clean immersive ambiance by the final mixer.

Stand seating is often limited with a large gathering at the eighteenth green and finish line of the equestrian event with roving crowd along the course. Each stand seating area should be captured with multiple mono and stereo microphones. Be mindful of the positioning around PA speakers.

The spectators at golf and equestrian events are reserved during actual play but are vocal and appreciative after a good performance. The problem is that microphones pick up the distant sounds of grandstand spectators reacting to something that has nothing to do with what is being presented. For example, the coverage is on the ninth tee and you may hear a reaction from a put on the eighteenth green. Note: this is a reason you should not have announcers outdoor on the eighteenth green or any other green.

Equestrian coverage is a lot like golf in the capture, mixing and playback workflow.

Cross-country equestrian and golf often require distant capture. Equestrian has many obstacles and jumps including water hazards that are covered with multiple spaced pairs and stereo microphones.

With golf there is PA and ice machines interference, however the PA is used conservatively and usually only to announce athletes. The Atmosphere perspective will not change during the production. The mixing perspective should be a looser impression with full 360 reproduction of action, reaction and ambient fill.

To further emphasize the action and its relationship in the sonic space elevating the tee microphones directly with an immersive sound mix fader or by bandpassing the high frequencies and elevating those specific frequencies further emphasizes the illusion of height. Off the tee the viewer sees the ball lift but capture is beyond possible. Emphasized in all forward speakers – left, right, left height and right height and is an example of using the height speakers for effect gives an illusion of height.

Equestrian and golf present competition live in real time and replays as if real time from playback sources. Matching the background ambiance is challenging because of spatial acoustics between sections of the course and noise interference.

Mixing with 3D Mixing Consoles

Specific sound elements and microphones can be easily positioned in an immersive soundfield using XYZ panners similarly as the sound element can be easily positioned in surround soundfields with surround and stereo soundfields with panorama controls.

Mixing Immersive with Surround Busses: Microphone Double Bussing/ Double Microphone Placement

Ear Level Layer #1 – 5.1 Microphone –L and R, front and back
Ear Level Layer #2 – 5.1
Ear Level Layer #3 – 5.1 Camera microphones
Ear Level Layer #4 – 5.1 Atmosphere microphones –L, R, LS and RS.

Height Level Layer #1 – 5.1 Atmosphere #1 microphones – L and R, front and back
Height Level Layer #2 – 5.1 Atmosphere #2 microphones – height left, height right, height left surround and height right surround.
Height Level Layer #3 – 5.1 Atmosphere #3 microphones – side height left, side height right, side height left and side height right.

Note: The ear layers and the height level layers do not share microphones.

Case Study 10.11

Immersive Sound for Extreme Summer Sports: Skateboarding and BMX – Big Air, Vertical Jumps, Pipe and More

Production Philosophy for Immersive Sound

Summer Extreme Sports is a multi-discipline group of non-motorized wheeled sports – skateboarding, BMX bicycles and rollerblades, anything that roll down and up ramps, through half-pipe courses and other extreme hazards. ESPN has introduced motorized sports which are not examined in this case study.

Each Extreme Sports has an opportunity for advance and interesting immersive sound design because of the ability to place microphones close to the athletes and separate specific details in the sound design from general ambiance and noise.

Big Air is similar to ski jumping except the athlete transverses down a steep ramp on a skateboard, bike or skates not an icy slope on skis. The athlete gets airborne and lands an extended distance down the course then goes up a ramp on the far side of the landing zone. The Big Air ramp and Pipe course is wooden and closely covered with handheld cameras and contact microphones.

The specific sports sound zone at the top of the ramp is captured as well as capturing a top-down on-axis perspective of the audience which is some distance away.

Contact microphones along the underside and length wise of the wooden ramp along with mono shotguns along the sides of the landing area cover the complete horizontal and vertical movement of each zone. The microphones are physically spaced and can be spatially oriented to give a spread-out horizontal dimension to the sound and picture.

The Pipe event clearly show horizontal and vertical motion in the pipe using extensive contact microphones at the bottom and top of the pipe wooden course which are spatially panned to complement the camera angles. The Pipe events benefit from significant horizontal movement but is dependent on atmospheric and ambient capture and enhancement for stable immersive soundfield.

Figure 10.12 The side perspective of all Big Air events shows a pronounced vertical separation between the top of the ramp and the landing zone with easy separation of the extremely different sound zones

Additional events include Skateboard Extreme, which is usually over obstacles such as pipe-rails, stairs, steps and other hazards. Each obstacle has a unique sound and presence and these sounds can be individually captured and imaged vertically to enhance the height aspect.

The view of the athlete is often head to toe and the athlete can be animated and vocal during the routine which visually supports sports sounds in the height speakers. Distinct and separate sounds can be captured from each athletic zone and with spatial placement effectively implement the principles of **F**ront **V**ertical **S**oundfield **E**nhancements (discussed in Chapter 4) which correlates and reinforces the front – top to bottom aspects of the sound mix as well as the front left-right aspects of the soundfield.

Microphones

A dimensional ambiance can be derived from microphones in front of and above the different audience zones. A large grandstand and milling around area views the Big Air event and should be captured from the POV of the athletes at the top of the ramp looking down at the audience. This capture space gives a stable overview of the events and should be behind any PA clusters that amplify sound to the spectators.

Crowd microphones should be placed on stands in front of the audience at ground level to capture a present on-axis sound and along with microphones suspended over the crowd a dimensional crowd layering is possible simply from microphone placement.

However, a large, open multi-sport production suffers from ambient noise pollution.

The ambient sound includes sports, crowd, PA and the reflections of everything inside the hard surfaces of the venues.

Create a compact 4.0 immersive sound zone around the athlete which is blended with a spacious 4.0 landing zone perspective. Concentric quad zones – ear level and from above.

Unfortunately there usually is a substantially active PA announcer and loud PA system.

The sound perspective of each zone will stay fixed during the entire event. The microphone's physical positioning and arrangement effectively creates an immersive soundfield giving a height/sound aspect to the picture.

Miniature microphones are placed on each corner of the mat and routed to the left, right, left height and right height channel for a full-frontal experience from the landing of the athlete.

The running zone is covered by stereo shotguns on microphone stands, however they are spaced fairly wide because of the large area for the vaulter to prepare.

Competition Zones and Atmosphere Production

Mixing with 3D Mixing Consoles

Specific sound elements and microphones can be easily positioned in an immersive soundfield using XYZ panners similarly as the sound element can be easily positioned in surround soundfields with surround and stereo soundfields with panorama controls.

Mixing Immersive with Surround Busses: Microphone Double Bussing/ Double Microphone Placement

Ear Level Layer #1 – 5.1 Zone Microphones – L and R, front and back
Ear Level Layer #2 – 5.1 Zone microphones
Ear Level Layer #3 – 5.1 Camera microphones
Ear Level Layer #4 – 5.1 Atmosphere microphones – L, R, LS and RS.

Height Level Layer #1 – 5.1 Zone microphones – L and R, front and back
Height Level Layer #2 – 5.1 Atmosphere microphones – height left, height right,
height left surround and height right surround.
Height level layer #3 – 5.1 Atmosphere microphones –side
height left, side height right,
side height left and side height right.

Note: Ear layers 1 and height level layer #1 share microphones.

Summary

The height aspect of the sound design will stay constant even with dimensional changes in the mix accounting for left–right and front–back camera perspective and orientation.

An ambitious goal is to create a holistic soundscape when cutting between five different events.

Case Study 10.12

Immersive Sound for Field Sports: Football, American Football, College Football and Field Hockey

Production Philosophy for Immersive Sound

With any sport sound scheme minimum immersive can be as simple as the addition of ambiance and atmosphere into the height speakers. This modest supplement usually creates a sense of aural space for the picture.

World Football and American Football are similar because the crowds are very tribal with a lot of chanting and singing. There is a tendency to over capture the diffused crowd to create a more homogenous home crowd effect with less front detail. College football is unique because the band and student section bring a distinctive air and sound of youth and can be easily captured and reproduced using front focus techniques.

Football (Premier League and American Football) is visually presented predominantly from a high-side camera, also known as the play-by-play camera, which is interspersed with cuts to field cameras.

Football is the definitive examples of the use of overhead atmospheric enhancements for spatialization. Additionally, overhead atmospheric enhancements do not require any localization and can be accomplished on most 5.1 mixing desks. Note: 3D panning is helpful for precise localization, but bottom line – convincing immersive sound for most sports does not require precise localization.

Generally field sports have no "sports-relevant" sound in the vertical axis (above the listener), only ambiance and atmosphere and artificial embellishments are probably not appropriate. Create the complete "man-in-the-stands" soundfield where enhanced ambiance is entertaining enough of an embellishment for football.

Cheerleaders are present in American and College football and may be an opportunity to inject some spatialization into the mix with head-to-toe sound between plays. Remember, immersive sound does not need to be over the top to be effective.

Front Vertical Zone Enhancement

The band in college football is usually loud and entertaining and easily fills the frontal zone in an immersive sound mix. With a microphone placed in front of, to the sides of and as much above the band a fully immersive sound is created in front and above the listener/viewer.

Effects/Mix Automation

One of the greatest challenges for field sports has been consistent field of play sounds.

Access to the sidelines is critical but sideline sound effects for American Football have been inconsistent because of league and player rules – half of the field is blocked from broadcast microphone access severely restricting the sound from the field of play. Additionally, Football/Soccer suffers from inconsistency because of sound pollution – vuvuzelas and loud PA systems. Little is said about distractions, but of significance is missing cues and kicks from fatigue and disturbances.

John Madden's Football Video games was the first out and prompted the question from players, listeners and entertainment consumers: "Why can't you hear this level of sound on broadcast?" This prompted people from FOX Sports, particularly Fred Aldous, to push for wireless microphones on the players.

Fader/mix automation systems like "SALSA" and the Lawo "Kick" consistent field of play was almost impossible. SALA uses the field of play microphones to generate a real-time sound mix or as triggers to playback samples from a library file.

Samples offer a high level of fidelity because there is complete separation of sound elements and no background coloration from opening and closing microphones, but samples usually do not provide the local color and flavor or the referee whistles.

Atmosphere

A dimensional crowd is created from spatial separation, layering and purposeful placement of microphones. Most stadium designs have at least two or three tiers for seating, making dimensional crowd layering possible simply from microphone placement with stands and mounts. There is room between the field of play and the seating to place microphones on stands in front of the crowd.

The ambient sound includes sports, crowd, PA and the reflections off everything inside the venues.

The upper left and upper right microphones anchor the height aspect of the sound design and their XYZ position is static above the listener and does not change with camera cuts.

Immersive sound for sports does not always need to be physically or proximately correct. For example, place the cheerleader in the upper speakers, the height aspect of the sound design will stay constant.

Summing 5.1 Busses

> Ear Level Layer #1 – 5.1 Field microphones – L and R
> Ear Level Layer #2 – 5.1 Atmosphere microphones – L, R, LS and RS.
> Ear Level Layer #3 – 5.1 Band Mix – L and R
>
> Height Level Layer #1 – 5.1 Atmosphere microphones – height left, height right, height left surround and height right surround.
> Height Level Layer #2 – 5.1 Atmosphere microphones – side height left, side height right, side height left and side height right.
> Height Level Layer #3 – 5.1 front band mix – height left, height right

Summary

Some sound above the viewer/listener will usually create a sense of aural space for the 2D picture (even at 4K resolution). But sound designers are wrestling with the concepts of what sounds should be heard above the viewer when there is no obvious reason.

Field sports may not need over-the-top type production to satisfy the listener/viewer.

Case Study 10.13

Immersive Sound for Notre Dame College (NBC) Football

Karl Malone (NBCUniversal)[3]

The senior mixer A1 Doug Deems used the 2018 and 2019 season to learn how to use this new vocabulary to better bring the fan at home into the stadium.

AUDIO – Notre Dame College Football in HDR and Atmos September 14 to November 23 2019 – Second season of Notre Dame College football on NBC in 1080P/4K/HDR Atmos as a single stream production from NEP's ND6 OB Van.

A Dolby DP590 reference decoder is installed on ND6 (1RU w/ MADI I/O) to emulate object-based scenes. The DP590 will appear on the Calrec monitor bus. Calrec provides support for monitoring immersive signals from an external source. ("Immersive Monitoring" is available in the>Show Settings>General Settings.)

We will use the DP590 to switch between downmix (5.1 and Stereo). The Dolby DP590 in the truck and the DP591 (encoder) in EC have been configured to trim the heights of the .4 (5.1.4) by -12db all but eliminating them in the downmix to 5.1. This means that we can effectively route a 10-channel mix to both SDR and HDR productions.

The SDR takes the first six channels and throws away the heights. The HDR takes all 10 channels for the 5.1.4. We will have .4 height specific microphones which are not required for the 5.1 SDR production.

No Dolby or Linear Acoustic support personnel will be onsite although we will have Linear Acoustic support at Englewood Cliffs for the new "Immersive Soundfield Controller" or ISC. The New ISC Upmixer (Main and Backup) will be in ND6 to upmix both 2.0 and 3.0 content (Music/Edits pieces) to 5.1.4.

When the unit is operating in "automatic" mode, it continuously monitors the incoming channels for audio and automatically determines the channel configuration of the input signal and which upmix mode is appropriate. "Automatic" mode will upmix 2.0, 3.0 and 5.1 to 5.1.4.

Full configuration, control, and monitoring of the UPMAX ISC is performed using the unit's web-browser based interface, however configuration of the unit's IP address is possible via the front panel interface. The UPMAX ISC unit only supports one user account and password. The default password is 1234.

The New ISC Upmixer in EC will upmix 2.0 and 5.1 commercials to 5.1.4 and sit across the X/Y output from EC. Latency through the ISC has been reduced from 23ms to 11ms, meaning it is now practical to run audio through it in the truck. The Upmixer in ND6 essentially will take a 3.0 Edit and send Ch. 3 (voice to center as normal) and take everything below 8K (indicated as "Front-High Crossover") from Channels 1&2 and upmix to the rear Ls/Rs. Anything above 8K will be upmixed by the heights. That crossover freq. (Hz) is moveable.

This effectively means that when cutting the .4 for SDR we will lose anything 8K and above on the 5.1 mix. We have yet to see/hear the result. This may need to be adjusted "on the day".

ND6 (OB Van) will produce the 5.1 Standard Dynamic Range (SDR) show using 1080P/ High Dynamic Range (HDR) cameras feeding 1080i to 30 Rock for Commercial integration and transmission. ND6 will also be producing the 5.1.4 HDR show with 1080P/HDR cameras feeding Englewood Cliffs (EC) for Commercial integration (plus 5.1.4 upmix) and for composite encoding to Atmos. Vision will be scaled 1080P to 4K for Direct TV transmission.

There will be a halftime show only, which will come from Notre Dame through the truck. Edits will be 3.0, graphic elements will be 2.0. Music and graphics will be upmixed through the ISC. In ND6 the A1 will predominantly be mixing for the 5.1 show. The main care and attention will be to ensure the Stereo Downmix of the 5.1 is good.

Minimally we will have a dual Stereo pair (BP4029) for the +4 heights. Thoughts are to place these two stereo mics high on opposite sides of the stadium. The height mic placement hopefully puts the ND PA stack in the rear heights LRH/RRH and in phase!

The 5.1.4 feed will be routed PCM with the 1080P HDR feed on a separate Tx line to EC. OMNI/ISO mics placed on/near camera platforms to complement the height mix.

The 5.1+4 will not be downmixed to 5.1. The first 6 channels contain the 5.1 surround program. This program should go to EVS for playback. Truck is 8 channels wide, not 16.

The EC lab can monitor baseband audio plus Atmos decode off router as well as Direct TV. This worked well for last season and gives us immediate feedback of the off-air product. We will measure LKFS loudness using the Dolby DP590 GUI. Tests proved consistent with what EC were monitoring (after Atmos encode).

What is in the 5.1+4? The 4.0 heights made up of *wide* stadium crowd effects via mounting 2 × BP4029 Stereo shotguns high on opposing roofs. Spot/Omni mics (ECMs) will be deployed in camera positions to add near crowd when appropriate to overheads. ND & Visiting bands. 1 × VP88 ND band in LRH and LFH- 1 × VP99 visiting Band in RFH and RRH. Referee mics, primarily panned to Center only for 5.1 SDR feed and added to LFH and RFH for Atmos. Painter poles for near Crowd L/R in 5.1 plane.

ST Fiber on roof at both ends. NBC prov. 2 × MP2 stereo mic preamps and 2 × Sescom FA2 sets Fiber/Analog to get two stereo mics to truck.

Case Study 10.14

Immersive Sound for European Football: BT Sports

Production Philosophy for Immersive Sound

The BBC struggled to keep the rights to televise Premier League and Championship Football and virtually all sports unless required by an entity like the IOC (International Olympic Committee) which requires free-to-air and terrestrial distribution which is available to most consumers, struggle financially.

Even before streaming changed the business model, rights fees became competitively bid up as private broadcasters expanded viewers and infrastructure as fiber and satellite developed. As streaming becomes the preferred choice (and means) of content distribution, the BBC's sports production was doomed.

British Telecom founded a unique high-quality production that merged well with a high-usage business model.

Jamie Hindhaugh, Chief Operating Officer for BT Sport and BT TV, believes audio is overlooked because people focus on picture quality. Now sports fans will be able to feel like they're in the stadium, with the combination of Dolby Atmos and UHD providing the most immersive experience possible.

I do not think anyone would deny that the BT business model is about selling bandwidth and getting a ROI on their infrastructure and content investment and there is no doubt that data streaming is the foundation for commerce and entertainment.

4K UHD with immersive sound uses more and needs faster bandwidth than your typical internet surfer and email user. BT along with other global players like AT&T not only have the conduit to the public but have heavily invested in content and committed to a higher level of production value.

British Telecom launched streaming sports in 2015 and for the 2019/2020 season BT rebranded and improved its service as BT Sports Ultimate 4K Ultra HD with 50 frames per second and Dolby Atmos sound. Sports broadcasts on BT Sports Ultimate include Premier League, Champions League and Europa League football matches, plus rugby, UFC, WWE and boxing. BT Sports requires a minimum of 25Mbps and there is no doubt that BT Sports looked at immersive sound as a premium attraction to its 4K UHD video content. Immersive sound and Dolby Atmos is defined by BT Sports as 5.1.4.

Telegenics provided the OB Van where Richard Williams served as the audio guarantee.

The workflow was organized around the mixing desks 5.1–buss structure which meant that the audio mixer didn't need to do anything different on the console for mixing immersive than for mixing 5.1 surround.

Calrec Apollo consoles, use one Main Bus 5.1 Output as the bed, another 5.1 Output for the 4 height channels (C & Lfe channels not used). The Apollo is configured for 16 Main 5.1 Outputs.

Sound design – a soundfield 1st order ambisonic microphone is the foundation for the immersive soundfield. Ian Rosem was responsible for the immersive sound design for BT Sports and told me that "more than half the football stadium across the UK have a Soundfield microphone already installed." Ian has worked in the football stadiums for decades and noted that "the Soundfield Microphone and position is basically the same as before and that only a DSF-3 MKII decoder was added to be able to decoded a variety of immersive sound formats from the single point source microphone."

The SoundField DSF-3 MKII digital surround processor supports a new range of object-based audio formats, in addition to Stereo & 5.1. These new formats include height information, 5.1+2, 5.1+4 & 7.1+2. So the bed is the original 5.1 with the additional high fronts and rears. So L. R. C. Lfe. Ls, Rs as per TV Dolby spec. Then Height channels FL, FR, RL, RR.

The FOP (field of play) is covered with 12 or 13 shotgun microphones depending on the ground and technical areas and where the touchline camera 3 is. One behind each goal, one on each corner and therefore three or four then spaced equally along the touchlines, including the one on Cam3.

Figure 10.13 Football kick and net microphones wide

Objects are usually used for commentators and stadium P.A., while FoP FX mics are mixed into the L and R, in mono, of the 5.1 bed. No doubt more objects will be added. I know the technology is to give the end listener the ability to "render" the audio how they wish. You could obviously have other commentaries as objects, more personalized fan experience using positional mics etc.

The sound mixer still provided a stereo and 5.1 mix for international distribution as well as a Dolby Atmos from one mixing desk. Stereo and 5.1 are sent back over fiber in a SDI format. The Atmos mix is encoded into a Dolby ED2 stream that carries the object metadata as well as the audio, then transcoded to Dolby Digital Plus and routed through the broadcast facility.

The stream is sent to consumers by BT's Infinity broadband fiber network using HEVC 4K encoding. Dolby Digital Plus does not support personalization – Phase 2 implementation will use AC-4 and include loudness controls and accessibility options, interactivity?

The viewer/listener must have a 4K BT Ultra Set Top Box, 4K a UHD television to see the difference and a Dolby Atmos compatible amplifier or soundbar to hear immersive.

25Mbps is minimum for 4KUHD with Dolby Atmos, BT's superfast fibre has an average speed of 50Mbps – note 70 Mbps is necessary for 8K HDR service.

BT is working on a gigabit broadband service which would exceed 1,000 Mbps. BT has tested 8K television with Samsung.

BT Sports content – Premier and Champions League Football, Boxing – Fight Night in Dolby Atmos. Dolby Atmos is backwards compatibility because phase one rollout was over Dolby Digital. BT Sports partnered with Samsung for 8K trials and it is expected that Samsung will be the first to support BT Streaming apps.[4]

Case Study 10.15

Immersive Sound for Gymnastics

Introduction

Gymnastics is a multi-discipline group of events with separate men's and women's competitions. During early stages of competition all apparatuses have simultaneous competition which presents unique challenges for audio capture and production.

All events except for floor exercise use athletic apparatuses to conduct and perform a routine. Fixed apparatuses include bars, balance beam, pommel horse, rings, vaults and a specifically prepared area for floor exercises. Additionally the entire competition space is elevated on a wooden floor which amplifies and isolates the runs and landings sound.

Each apparatus and event provide distinctly separate sound zones that can be mixed and positioned in an entertaining and interesting dimensional soundscape.

Production Philosophy for Immersive Sound

Gymnastics presents the opportunity for advanced and interesting immersive sound design because of the ability to place microphones close to the athletes and separate specific details in the sound design from general ambiance and noise. There are distinct and separate sounds that can be captured from each apparatus and event. Additionally, these sounds are visually associated with different parts of the field of play, including the apparatus, runways and landing matts.

The gymnastics immersive sound design effectively uses the principles of Front Vertical Soundfield Enhancements (discussed in Chapter 4) which correlates and reinforces the front – top to bottom aspects of the sound mix as well as the front left-right aspects of the

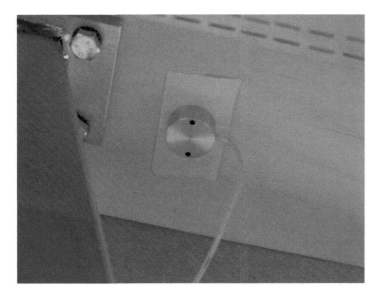

Figure 10.14 Contact microphone are fixed to the bottom of the balance beam and the pommel horse capture the athletes' contact with the beam and the horse without adding acoustic/ambient noise to the mix

soundfield. Essentially the sound is dispersed into the entire anterior vertical plane of the viewer/listener.

For example, from the side perspective of the women's uneven bars the athlete appears as though they are above and below eye/ear level during the routine. Microphones positioned on the apparatus and floor are physically spaced and can be spatially oriented to give a spread-out vertical dimension to the sound and picture.

Men's rings are visually very vertical and provide positions to mount the microphones above and below the athlete during the routine.

Men's and women's vaults not only have a pronounced horizontal movement but also some vertical separation between the run-up, spring, vault touch and landing.

The balance beam and pommel horse have vertical separation between the apparatus and mat.

The women's floor exercise is unique because it is a cross between aerobics and dance and utilizes music. The music should be up-produced and used along with the atmosphere to provide a cohesive dimensional soundfield. The men's floor exercise does not use music and is the most challenging to find separation between the floor and room.

A dimensional ambiance can be derived from microphones in front of and above the audience plus a subtle blend of PA.

Even though there are several different visual perspectives, the sound perspective will stay fixed.

Microphones

Many of the microphones can be placed around, below and above the athlete and by their physical positioning and arrangement effectively create an immersive soundfield giving a height/sound aspect to the picture. Close microphone placement is possible for gymnastics on all athletic apparatuses using low-profile directional and contact microphones.

Figure 10.15 Gymnastics boundary microphone

Production Note

The Gymnastics Federation will want to approve the microphone placement and rigging and ultimately has the final say on where a microphone is placed and all issues and decisions on the field of play. Apparatus microphones and cables must be constantly scrutinized for safety and appearance.

Gymnastics is covered visually with close camera perspectives; however, miniature microphones can be skillfully hidden from camera view. Miniature microphones are secured at the flex point of the wooden bar and on the metal framing of the apparatus.

The athlete's movement, running and jumping creates an acoustically audible sound above the podium, however these sounds can be captured in the space under the podium acoustically with diaphragm microphones or through the vibrations in the floor with contact microphones.

Apparatus Zone and Atmosphere Production

Gymnastics is unique because of the large field of play footprint, with up to five apparatuses active at a time with simultaneous competition and coverage. Each apparatus is its own production zone with specific apparatus and localized atmosphere sound.

A dimensional crowd is created from spatial separation, layering and purposeful placement of microphones.

Many venue designs have at least two or three tiers for seating that are close to a particular active area on the field of play, plus there is usually a space between the field of play and the seating areas which simplifies crowd microphone placement. Crowd microphones should be placed on stands in this space to capture a present on–axis sound and along with microphones suspended over the crowd a dimensional crowd layering is possible simply from microphone placement.

However, a large, open multi-sport production suffers from ambient noise pollution.

The ambient sound includes sports, crowd, PA and the reflections of everything inside the hard surfaces of the venues.

Mixing with 3D Mixing Consoles

Specific sound elements and microphones can be easily positioned in an immersive soundfield using XYZ panners similarly as the sound element can be easily positioned in surround soundfields with surround and stereo soundfields with panorama controls.

Mixing Immersive with surround busses: microphone double Bussing/ Double Microphone Placement

Ear Level Layer #1 – 5.1 Apparatus microphones – L and R, front and back
Ear Level Layer #2 – 5.1 Floor microphones
Ear Level Layer #3 – 5.1 Camera microphones
Ear Level Layer #4 – 5.1 Atmosphere microphones – L, R, LS and RS.

Height Level Layer #1 – 5.1 Apparatus microphones – L and R, front and back
Height Level Layer #2 – 5.1 Atmosphere microphones – height left, height right, height left surround and height right surround.
Height Level Layer #3 – 5.1   Atmosphere microphones – side height left, side height right, side height left and side height right.

Note: Ear Layers 1 and height level layer #1 share microphones.

Summary

An aspirational goal is to create a holistic soundscape when cutting between five different events. The height aspect of the sound design will stay constant even with dimensional changes in the mix accounting for left-right and front-back camera perspective and orientation.

Case Study 10.16

Immersive Sound for Handball

Production Philosophy for Immersive Sound

Handball is a sport that has been presented in full surround with sound typically dominant in the front, left and right as with most stereo productions. This stereo sound design is still good practice for the ear level soundfield in surround and in the immersive mix.

Generally, handball has no "sports-relevant" sound in the vertical axis (above the listener) with one visual exception – the close-up of the goal and goalkeeper. The goal can be easily captured dimensionally, which can be used for frontal vertical enhancement with the goal sound radiating from top to bottom.

There should be four miniature mono microphones along the left, top and right side of the goal and stereo microphones on the handheld cameras. There is sufficient physical separation between the four field of play microphones on the goal to create an effective front vertical soundfield.

The court sound is similar to basketball and can be captured with microphones along the side of the court or with microphone operators. Boundary microphones are effective because they can be placed close to the action on the field of play and be safe for the athletes. Court microphones should be spatially enhanced through spatial expansion (see plug-ins) of the stereo and mono microphones. Algorithmic spatial expansion fills any dead spots in the immersive soundfield and is not used for localization.

There is clear separation between seating and the field of play and room to place microphones on stands in front of the audience. A combination of microphones in front of the seating stands as well as over and even behind the audience for some off-presence atmosphere contributes to the immersive sound experience.

Overhead atmospheric enhancements can be accomplished on most 5.1 mixing desks. Note: 3D panning is helpful for precise localization, but the bottom line – convincing immersive sound for most sports does not require precise localization.

This sound design creates the complete "man-in-the-stands" soundfield where enhanced ambiance and a focused sports action sound is entertaining enough of an embellishment for this sport. Immersive sound does not need to be over the top to be effective.

Sonic frequency expansion and band splitting is possible by routing microphones and combinations of microphones to separate surround or immersive group busses. Remember the Blauert effect and the perception of space between high and low frequencies.

Case Study 10.17

Immersive Sound for Indoor Ice Sports: Figure Skating, Hockey, Speed Skating and Curling

Creating perspective with sound design is how we want to focus the attention of the viewer and is a significant consideration that went into the sound design of every sport.

Indoor ice venues for skating sports derive their soundscape from sounds that are at ice level, such as the slicing of the skates and sounds that are above ear level, such as the ambiance of the venue. Indoor venues have a mental image because of the expected reverberant field.

Visual coverage is usually presented from camera angles looking down on the field of play and at ice level. From the camera and spectator point of view there is no separation in elevation of the sounds from the field of play and no visual reason to elevate sports sounds such as skates or impact sounds.

The perspective is defined by the use of the atmosphere, sports specific and venue sounds to set the mental image and expectations of the viewer.

Immersive Sound Philosophy for Production of Ice Sports in Indoor Venues

Indoor immersive sound design correlates all aspects of the sound elements and soundfield to create a natural immersive sound design. The immersion sound experience for indoor ice sports is connected with the venue acoustics, which is an integral aspect of the immersive sound design for these sports. An indoor venue has a distinct room tone that is contoured/shaped by the acoustic characteristics of that venue.

The perspective of hockey and the skating events is typical of a person in the stands viewing the action in front of them and appropriate for many productions. The viewer's attention is focused forward to the picture where the effects mix is contained in the front speakers and the viewer is surrounded by atmosphere from the front and rear speakers.

Hockey and speed skating uses close microphones to capture the effects at ice level and reduce the room tone in the front left and right speakers/channels.

Hockey is unique because there is a barrier that partially shields the microphones from crowd and PA. A consequence of this barrier is an amplification of the skates and stick sound from the reflected waves in the bowl. Often there is a PA cluster over the center of the ice,

Figure 10.16 Setting up boundary microphones along the ice in figure skating

along with the possibility of more distributed speakers along the walkway above the spectators. Essentially all PA speakers are above the venue seating and often the PA is very loud and overused.

Capture the space – I have placed large diaphragm mono and stereo microphones around the venue to capture the dimensional tone of the venue. The high frequencies from the crowd and PA exaggerated in the reverberant field blend to form a natural dimensional soundfield similarly to being in the space.

The LFE is not used for bass management and very few sports will generate sound for the LFE. The corner boards in Hockey was one of a few instances in Winter Sports where use of the LFE was effective.

After the start, the sound of Speed Skating is dominated by crowd and PA. Capture the room with mono and stereo microphones and construct the dimensional space by placing the sounds in the soundfield – recreating the soundspace.

The sound of figure skating is the combination of skates, music and audience engagement that creates a unique atmosphere and tone of the venue and should be presented in surround sound as to encircle the viewer.

At figure skating the music is played into the venue at very loud levels and the reverberant field of the music is picked up in the ambiance and atmosphere microphones and blended into the mix.

Good levels of the direct music signal bring the music up front in the mix and give presence to the reverberant field.

Reverb and room simulator tools should be used for spatial expansion of the music and give a stable immersive soundfield to be balanced against the acoustic reproduction of the music tracks. The expanded musical tracks should be used to separate and give definition to the channels while the room reproduction provides the expansiveness of the event. This method gives the sound designer the ability to vary the ratio of the dry, expanded signal and wet acoustic signal.

The music in figure skating can end up in the LFE channel through up-processing and specific microphones can be band-passed and routed to the LFE channel.

Figure skating provides the opportunity for completely changing the expanse of the sound zone in the competition area as compared to the soundspace of the "Kiss and Cry" zone which needs a more intimate feel. The picture from Kiss and Cry is generally a medium chest and head shot with athlete and coaches sound filling the front top to bottom sound zone with the venue background behind the viewer/listener.

Microphone Placement

Immersive sound design for indoor ice sports is built on microphone placement to capture an acoustically detailed soundspace and conveying specific sound details and their relationship in the sonic space – capture the space.

Many venue designs for speed skating are temporary and have space between the field of play and seating where microphones can be placed on stands in front of the crowd. It is always desirable to capture crowd sound on-axis from the front.

Temporary tiers for seating provide spots to mount microphones and capture dimensional crowd layers simply from microphone placement. Be conscious of PA speakers and try to position the microphones off-axis to the speaker.

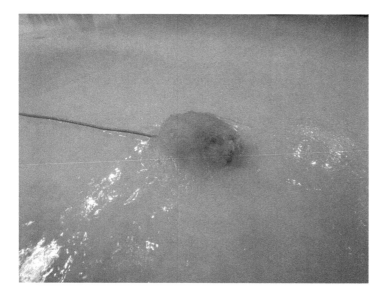

Figure 10.17 Microphones in ice

Mixing

The balance between close, mid and diffused layers of sound acquisition is the foundation for an indoor venues soundscape.

Mixing comments Simon Christian Tonemeister Fraunhofer Institute include:

> Figure Skating 100% hard panning of each on the mid and upper layer perhaps would generate an impression that the ambiences are not associated to each other. I would try a slight mix of both to the other layer.
>
> Another thought is that perhaps a slight orientation to the front channels would guarantee a powerful impression of the music, which sometimes is lost by distributing the energy on too many channels.
>
> The PA announcer could be a good signal to pan it in the upper layer, if this fits with the natural impression of a PA under the roof. This could "widen" the mix and leave space for the nearfield mics. A dynamic mix between stadium ambience and music according to the action on the ice is desirable.
>
> There are mics over the field of play in ice hockey in addition to the boundary mics – its desirable to give them more space in the 3D mix to be more in the field action.[5]

Mixing with 3D Mixing Consoles

Specific sound elements and microphones can be easily positioned in an immersive soundfield using XYZ panners similarly as the sound element can be easily positioned in surround soundfields with surround and stereo soundfields with panorama controls.

Mixing Immersive with Surround Busses: Microphone Double Bussing/ Double Microphone Placement

Ear Level Layer #1 – 5.1 L and R, front and back
Ear Level Layer #2 – 5.1 Camera microphones
Ear Level Layer #3 – 5.1 Atmosphere microphones – L, R, LS and RS.

Height Level Layer #1 – 5.1 Microphones – L and R, front and back
Height Level Layer #2 – 5.1 Atmosphere microphones – height left, height right, height left surround and height right surround.
Height Level Layer #3 – 5.1   Atmosphere microphones – side height left, side height right, side height left and side height right.

Note: Ear layers 1 and height level layers do not share microphones.

Case Study 10.18

Immersive Sound for Figure Skating: 180–degree VR Camera Production

Let's look at figure skating where I will define the basic sound design goal is to extend the immersive experience into the field of play and less of a user-controlled experience.

A dimensional experience works well with 180-degree video because these cameras are optimized for forward resolution and the need for much, if any, head tracking is decided on early. The VR production was accomplished using four 180-degree camera pods. Two camera pods at ice level in opposite corners of each other, one medium wide POV and one very wide POV.

Intel Sports injects an interactive element into their production because the viewer/listener can switch between each camera pod. The panorama view of each camera is about the same as most people's peripheral vision, which means there does not need to be a dramatic head move to see an already wide front panorama and unless the viewer/listener is cued there may be no need to turn their head completely around to see what is behind them unless there is a danger sound clue.

First, it is essential to capture the audio from the POV of the camera. The cost of a 1st Order Sennheiser Ambisonic Microphones is insignificant (US$1,000) compared to the significant benefit for a 3D sound print of the camera zone. The audio serves as a localized sonic reference to that specific picture. Since Intel offers an option to select a camera it is critical to have local audio that matches that camera POV. Ideally this camera microphone position should be ambisonic, which would allow for unrestricted 360-degree audio of that camera position – forward field of play and around and behind is the audience. Certainly the sound perspective will shift from the highlighted sports sound from the field of play with the audience sound off-axis to where the audience is present and on-axis and the field of play is off-axis. Ambisonic allows for that; with every degree of head rotation there is a proper representation of the soundfield that matches the picture.

With all international competition there is an entity called the host broadcaster that produces core sound and picture coverage for the rights holders of the competition. With all microphone capture there can be limitation and this certainly is the case with low resolution ambisonics microphones. Additional microphones (spot microphones) should be added to the ambisonics mix from each camera position. For example, OBS does a thorough job of detailed sound capture and has dozens of microphones around the ice and in the venue which can be isolated and provided to the VR production team. Localization can be captured or created. Captured sound from microphones presents the problems of having a strong enough audio signal above the noise to adequately give good directional cues.

The immersive sound mix from each on-the-ice camera position can infuse very specific spot microphones from that location in the frontal soundspace, that heightens the effect of the skater skating close and by the camera pod. This sound can be mixed dimensionally enough where head tracking is not necessary since the illusion of the sound coming toward and leaving the camera creates the effect.

The natural immersive atmosphere at figure skating is a combination of audience, PA and music. There are some sound elements that you want to have a stable position. For example, a voice moving around could be unsettling. Any audio announcers or dialog tracks should be created as objects with user control so they can be removed if desired. A producer may be reluctant to use this type of feature so remember there can be parameter restriction built into the metadata.

Case Study 10.19

Immersive Sound for the National Hot Rod Association (NHRA)

Production Philosophy for Immersive Sound

Over The Top (OTT) is streaming with various data and quality limitations immersive sound since 2018 using a Calrec 8 channel buss configuration.

Generally, only the audio mix positions will need to be fit for 5.1.4 immersive sound monitoring. In the US, Game Creek and NEP are OB providers who have modified existing OB vans for immersive sound mixing.

For the National Hot Rod Association (NHRA), drag racing, Game Creek installed similar speaker setups in both mix rooms for A1, Josh Daniels and effects mixer, Rusty Roark to listen to. I was told that the mix rooms were small and there had to be compromises with the positioning of the speakers. Before my visit, I asked Mike Rokosa with NHRA to send me a 5.1.4 production clip to preview on my 11.1 Genelec system. In the studio, I was impressed with the movement of the sound in the horizontal and vertical soundfield and even more amazed after I personally spent a weekend at the drag races. I sat in the A1 mix position and listened to a couple of races. I could clearly hear the intricacies of the immersive effect mix in the OB van. I admit I was surprised at the positioning of the height speakers but did not let my eyes prejudice what my ears heard, particularly since I had already auditioned the mix in an accurate monitoring environment.

Remember, most OB vans have limited space to mount more speakers, much less accurately align and calibrate them. I spoke to Thomas Lund at Genelec Speakers about this problem and he said that compact speaker designs are available and Genelec's 8331 systems has been installed in applications with a listening distance of as little as 70 cm, which still delivered an accurate representation of the soundfield to the mixer.

I also talked to Mike Babbitt about tuning the speakers in the Game Creek OB mix rooms. Mike is the applications engineer from Dolby who worked with NHRA on the implementation of immersive sound. He told me,

> All speakers (not the sub) were set to reproduce the same SPL at the listening position using pink noise. For the sub, we try to set it with 10dB in-band gain compared to the center channel, which approximates to about a 3dB SPL bump using an SPL meter.

There was no time alignment of the speakers.

Another compromising factor is that the immersive 5.1.4 mix is being accomplished on concentric 5.1 busses. Although I am a big advocate of true XYZ panning, all I can say is that the NHRA sound model is very creative, clever and it works for their sport. Watch for my interviews with Josh and Rusty detailing the audio mapping and mixing that makes this an artistically creative mix.

As immersive sound production develops, we should consider alternative mixing solutions for live sports and entertainment. Mixing the immersive sound (surround sound) from a different location has been successfully accomplished for several World Cup productions. Master control should be monitoring immersive sound and could easily composite the audio stems from the OB van and build an immersive mix in a centralized control room. This is clearly a solution for many field-type sports such as baseball, football and soccer (the other football).

One final consideration: I suggest researching virtual/programmable soundfield reproduction for the audio mix room. Beam steering of microphones and speakers to create virtual soundfield patterns is a developed and well-understood technology and, as critical listening space becomes a necessity, perhaps an alternative approach to monitoring in the OB van should be assessed.

There will be a variety of immersive sound productions that requires a range of immersive sound monitoring. Critical listening for localization may be far more important in film, but when the height channels are a wash of distant crowd and PA, then I would suggest that for many sports (not all) critical listening is more of a judgement about spatialization of the sound elements than localization. Also, consider that the soundbar is the probable method of delivery for immersive sound and that precise localization from up-firing and side-firing speaker configurations is not possible in most home applications.

Downmixing Immersive to Other Formats

Downmix produced in OB Van during live production.

Case Study 10.20

Immersive Sound for Squash and Racketball

Production Philosophy for Immersive Sound

The sport of squash is presented from inside the enclosed court from the athlete point of view blended with a lively audience and "The Voice of God from the umpire that overlooks the game court and competition".

Squash is streamed on Squash TV and receives broadcast coverage every four years.

Squash and racketball both have a relatively small athlete field of plays and the sports action is confined and relatively easy to capture.

There is "sports-relevant" sound in the vertical axis above the listener and the camera perspective emphasizes spatial enhancements. The sport is presented from low camera angles where the athlete and racket are often seen head to toe. There are three distinct sounds – the running/squeaks, the slap of the ball by the racket, and the sound of the ball bouncing off the glass all resonating 360 degrees around the fishbowl.

The soundfield can be presented from inside the fishbowl, the athletes POV and outside the bowl from the spectator.

Court-side seating is close and entertaining with the competition appearing front and center – top to bottom from a listener seating at court side. The spectators also appear in the background through the clear plexiglass.

The spectators POV reinforces the use of front vertical enhancements – top to bottom sound design and atmospheric enhancements. From the athlete's POV, inside the bowl, the sound is completely different, full of lush echoes and reverberation and the athletes really do not hear the crowd well inside the walls even though the top is open.

There is clear spatial and sonic separation between the athlete and foot sounds on a wooden floor and the percussive nature of the ball slaps. The foot and competition sounds can be bandpassed, separated and mixed to enhance the spaciousness of the picture.

The referee sounds come through the PA, however the referee position is high above the court and the voice often seems to have a voice of God presence.

Microphones

The plexi-glass wall and wooden floor are the dais for the capture of a dimensionally diverse and lush set of sounds. The desire is to have microphones in the reverberant field, microphones outside of the reverberant field that will obviously capture the reverberant soundfield and transducers that do not capture the airborne reverberant soundfield.

There should be four mono microphones mounted on the walls, outside of direct sound reflections aimed at the field of play. There are miniature microphones inside the bowl front, center and right along the board below the score zone, but outside of the field of play to capture the acoustic soundfield in the bowl. Additionally, there are two more miniature microphones in the rear corners halfway between the floor and the top of the glass. Finally microphones attached to and under the table contribute to a wide range of sound that is isolated and can be used for creating a dimensional soundfield.

Mixing

Even though there is a thick reverberant/flutter echoes soundfield in the bowl directly processing some microphones through sonic expansion processes will give the sound designer a wide range of tones to work with to balance the most pleasing and interesting soundfield for the listener.

Sonic frequency expansion and band splitting is possible by routing microphones and combinations of microphones to separate surround or immersive group busses and pan and separate the high frequencies into the upper space and the low frequencies and ear level and the LFE. Remember the precedence effect (Blauert) of high and low frequencies.

Spatial expansion (see plug-ins) of the stereo and mono microphones by room simulators and other algorithms for dimensional soundfield enhancements into the height speakers creates an intimate sense of aural space for the picture.

With this style of sound design, algorithmic spatial expansion fills any dead spots in the immersive soundfield and is not used for localization.

Ambiance and Atmosphere

Floor-side seating at major squash events provide an opportunities for interesting immersive sound creation from microphone placement, capture and spatialization practices. A combination of microphones in front of the seating stands as well as over and even behind the audience for some off-presence atmosphere contributes to the immersive sound experience. Overhead atmospheric enhancements do not require any localization and can be accomplished on most 5.1 mixing desks. Note: 3D panning is helpful for precise localization, but the bottom line – convincing immersive sound for most sports does not require precise localization.

Mixing: The Voice of God

A focused field of play with a clear separation from the audience ambiance creates the complete "man-in-the-stands" soundfield with enhanced ambiance and a focused sports action sound. Immersive sound does not need to be over the top to be effective.

Case Study 10.21

Immersive Sound for Tennis and Table Tennis

Production Philosophy for Immersive Sound

The field of play at tennis and table tennis provides room for good microphone positions for progressive and cinematic film-style immersive sound design. Suspend belief and enjoy the new reality.

Tennis action is unique because visual coverage uses a high over-the-shoulder camera perspective to view both sides of the net creating a visually exaggerated front-to-back perspective. Tennis has distinct sound variations between the serve and the play action – the two focus points of the sound design. The serve tends to be the furthest from the net and stationary while the play action tends to get closer to the net and moves over virtually every piece of real estate possible to return a serve.

Figure 10.18 Table tennis net microphone

Even though there is minimum space (distance) between any microphones above the head of the athlete and the court surface, there are still options for spatialization up-processing methods to expand the spaciousness of the soundfield with room and space simulators.

This immersive sound design uses the principles of enhanced soundfield production, which emphasizes film-style production and reinforces the sound in front, behind, to the sides and above the listener/viewer.

The camera framing and perspective is constantly changing in tennis and because there is traditionally only one camera over only one player the serving perspective must change as well – front to rear and rear to front.

Front to back correlation emphasizes the surround sound aspects of the design placing the spectator on the court with the athletes.

Front to back correlation, as well as top to bottom correlation, can be problematic for summing surround and height channels into derivative mixes. As previously mentioned in Chapter 3, many broadcast consoles have built in surround to stereo algorithms with parameter controls.

The ambiance and atmosphere at most tennis events is reserved and naturally dynamic.

Microphones

Net sports have similar sound capture designs because microphones in the net are great locations to capture the sound. After the players serves they advance closer to the net and the microphone is at the center of the action, often very close to the athletes and usually the audio is on axis to the net microphone.

Microphones can be threaded into the net from each side (see Figure 20.19). Two microphones are ideal for most net sports. You want to place two microphones in the net, approximately one-third of the distance from each net pole. The two microphones in the net would be considered a spaced pair and with some panning to the left and right – 10 o'clock and 2 o'clock. This design will give a very full dimensional sound without conflicting with the commentary, which is in the hard center in surround or the phantom center in stereo.

You must consider that there is a danger of the ball or the athlete hitting the microphones. Of course, but some judicious placement will generally minimize this possibility. Placing the net microphones at the bottom of the net will usually reduce the chances of the microphone being hit by the ball. The microphone will be placed at the top far left and right corner of the net.

The upper left and upper right microphones anchor the frontal height aspect of the sound design and their XYZ position is static above the listener and does not change with camera cuts.

Tennis has a large field of play area between the service zone and the net.

Tennis benefits from the addition of floor microphones. For example, boundary microphones will clearly capture any feet and sneaker sounds and can be positioned mid-court at the net under the umpire's chair. Microphones mid-court give a sonic separation between the net and field of play sounds – racket and ball and vocalization of the athlete.

The serve can be covered by mono shotguns, for example the BP4071 or BP4073 aimed up the side boundaries or/and by a MS Microphone such as the AT4050ST in the middle of the backcourt. Mono shotguns aimed up the sides have been used for decades and are still in practice but there are mixers that have successfully used a MS microphone in the center of the back court.

It has been my experience to use caution when using these microphones when the athlete approaches the umpire chair because the language may not be ready for broadcast and could

Figure 10.19 Microphone pointing at the backline – service area at tennis

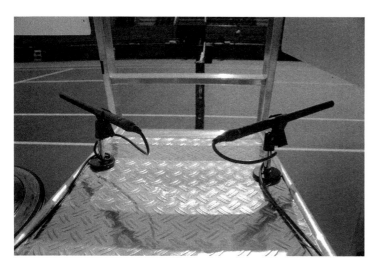

Figure 10.20 The umpire's chair is another useful place for microphones and helps to cover the serve from the mid-court area plus any conversation between the athlete and umpire

be considered private. Microphones should be aimed to the left and right as you are looking from the umpire's point of view, which easily covers both sides of the court.

Tennis is an interesting sport with generally smaller crowds that are close to the court and often quiet during the serve with usually only applause and cheering at the end of the action or score. With good microphone placement, tennis is a sport that almost mixes itself because of the dynamics of the audio.

In addition to the field-of-play microphones (back court and mid court), mini lapels near the athletes' rest area offer an excellent opportunity to sense the mood of the athlete and to pick up strategy during doubles competition.

Ambisonic Microphone Coverage

Tennis is one of the few sports that could be covered by a single ambisonic microphone given proper placement. The sound of tennis is very segmented. There is a general ambiance as the athlete prepares to serve and it becomes quiet before the serve. There is little crowd reaction after the serve and during the competition, only at the end of play.

A single 4th Order ambisonic microphone on each side of the net would be able to capture the complete field of play and entire atmosphere, from close to diffused.

Microphone Double Bussing/Double Microphone Placement

Ear Level Layer #1 – 5.1 Net microphones – L and R
Ear Level Layer #2 – 5.1 Serve microphones – left and right game camera
Ear Level Layer #3 – 5.1 Serve microphones – front and back Rear POV Camera Y
Ear Level Layer #4 – 5.1 Serve microphones – front and back Rear POV Camera Z
Ear Level Layer #5 – 5.1 Atmosphere microphones – L, R, LS and RS.

Height Level Layer #1 – 5.1 Net microphones – height left and height right
Height Level Layer #2 – 5.1 Atmosphere microphones – height left, height right, height left surround and height right surround.

Height Level Layer #3 – 5.1 Atmosphere microphones – side height left, side height right, side height left and side height right.

Note: Ear Layers 1 and Height Level Layer #1 share microphones.

Atmosphere

A dimensional crowd is created from spatial separation, layering and purposeful placement of microphones. There is room to place microphones in front of the crowd, including behind the umpire chair and along the barrier on each side between the field of play and the crowd lines. Often there is space in the corner areas for microphones on stands.

Wimbledon is a case study unto itself with an open and close roof on Court One. Not only is the sound of rain against the roof annoying, but the closed roof significantly changes the acoustics of the venue.

Summary

The height aspect of the sound design will stay constant even with dimensional changes in the mix accounting for left–right and front–back camera perspective and orientation.

The ambient sound includes sports, crowd, PA and the reflections of everything inside the hard surfaces of the venues.

Case Study 10.22

Immersive Sound for Volleyball and Beach Volleyball

Immersive Sound Production Philosophy

With any sport scheme minimum immersive sound design can be as simple as the addition of ambiance and atmosphere into the height speakers, which usually creates a sense of aural space for the picture.

Volleyball presents the opportunity for advance and interesting immersive sound design because of the simple camera coverage and ability to place microphones close to the athletes. Volleyball action predominately uses a high-wide camera perspective to view both sides of the net. The angle and framing of this picture shows the net and umpire clearly higher than the athlete.

This immersive sound design uses the principles of frontal vertical soundfield enhancements (discussed in Chapter 4) which emphasizes and reinforces the front – top to bottom aspects of the mix.

The immersive sound design for net sports such as volleyball and beach volleyball is built on the positioning of multiple mono microphones in the net. Net microphones have traditionally been placed along the bottom of the net to avoid hands touching the microphones, however additional microphone will be placed at the top far left and right corner of the net. The upper left and upper right microphones anchor the frontal height aspect of the sound design and their XYZ position is static above the listener and does not change with camera cuts.

Indoor volleyball can benefit from the addition of floor microphones. For example, boundary and contact transducers will clearly give a sonic separation between the net, hands and vocals of the athlete counter to the ball plus jumping and landing sounds against the floor.

Microphones have been placed in the sand at beach volleyball but the resulting sound is higher in pitch than the brain is expecting and does not provide the same level of sonic separation as the sound against a wooden floor.

Volleyball and beach volleyball have a large field of play area between the service zone and the net. Microphones can be placed at the umpire chair *photo* pointing at the backline (service area).

The camera framing and perspective is constantly changing in volleyball and beach volleyball. Left–right orientation is fundamental to the game camera, however volleyball and beach volleyball switch often to a near-far orientation with the camera view over the shoulder of one of the athletes.

Front to back correlation emphasizes the surround sound aspects of the design, placing the spectator on the court with the athletes.

Microphone Double Bussing/Double Microphone Placement

Ear Level Layer #1 – 5.1 Net microphones – L and R
Ear Level Layer #2 – 5.1 Serve microphones – left and right game camera
Ear Level Layer #3 – 5.1 Serve microphones – front and back rear POV camera Y
Ear Level Layer #4 – 5.1 Serve microphones – front and back rear POV camera Z
Ear Level Layer #5 – 5.1 Atmosphere microphones – L, R, LS and RS.

Height Level Layer #1 – 5.1 Net microphones – height left and height right
Height Level Layer #2 – 5.1 Atmosphere microphones – height left, height right, height left surround and height right surround.
Height Level Layer #3 – 5.1 Atmosphere microphones – side height left, side height right, side height left and side height right.

Note: Ear Layers 2, 3 and 4 share microphones.

Atmosphere

A dimensional crowd is created from spatial separation, layering and purposeful placement of microphones.

There is space behind the umpire chair and stand and along the barrier between the field of play and the crowd lines to place microphones in front of the crowd. Often there is space in the corner areas for microphones on stands.

Beach volleyball is a party atmosphere and "Made-for-TV sound". Along with the dimensional audio aspects of the sports sound and atmosphere, the ambiance including music and DJ can be spatially arranged to enhance an immersive sound mix.

Summary

The height aspect of the sound design will stay constant even with dimensional changes in the mix accounting for left-right and front-back camera perspective and orientation.

The ambient sound includes sports, crowd, PA and the reflections of everything inside the hard surfaces of the venues.

Case Study 10.23

Immersive Sound for Outdoor Winter Sports: Downhill Skiing, Cross Country and Biathlon Skiing

Production Philosophy for Immersive Sound

The Alpine events are particularly difficult for immersive sound because on the course there is virtually no distinct sound with height information. The course is open and the trees are away from the field of play minimizing any wind and tree rustle.

Perspective sound design is how we want to focus the attention of the viewer at the start, on the course and at the finish. The alpine events will benefit from periods of time when the immersive effect is noticeable and periods of time when then sound effects are dominant.

Advance audio production should be looked at as an umbrella of tools that not only can adjust the spatial properties of sound elements, but also change the tone and sonic characteristics of spaces. Basic spatialization can be as simple as time and timbre difference between a direct sound and a delayed or diffused element of the original sound.

There are four stereo stationary microphones plus several cameras with stereo microphones that when fully combined and imaged significantly contribute to the immersive image in the start house.

This effect is particularly successful when the athlete leaves the start tent and moves down the mountain. There is a stark change in picture from inside the start tent to a series of visuals that shows the expanse of the mountain. This sound design has a significant change in sound from the start tent to the mountain.

Dynamic expansion from a subtle ambient soundfield to an expansive outdoor soundfield from a properly captured ambiance and atmosphere is the basis of many outdoor immersive sound designs. Moderate silence to jubilation is an exhilarating aural effect for immersive sound.

Downhill skiing and jumping emphasizes the sonic differences in the isolation at the top of the jump, start house or on the course to the finishing and landing zones which are immersive sound zones unto their selves. There is an artistic skill in blending the transition from "lack of sound" to an immersive soundfield.

The course sound effects are captured from an operated approach microphone and a camera microphone. Fill microphones and "playback samplers" are used on some long stretches between cameras. The microphones are often mono because they need reach to capture the sounds, however these microphones would benefit from additional processing to expand the horizontal and vertical soundspace around the listener.

Finish zone – as the skier moves down the course there is a subtle increase the atmosphere at the finish. The finish zone is defined and initially seen as the cameras track the athlete down the hill. The stands are steep, giving a reason to have top to bottom crowd fill. The immersive sound zone at the finish is dominant front top to bottom with the rear soundfield hearing the quietness of the mountain.

The sports sound of skiing changes the viewers' attention zone three times. Beginning in the start tent, the immersive sound is close as if you are standing next to the athlete – equally front to back and top to bottom. The "submarine effect" fully fills the soundfield.

On the course the sound is forward to the picture where the effects mix is contained in the front speaker's ear level and height speakers where the viewer is surrounded by atmosphere from the front and rear speakers.

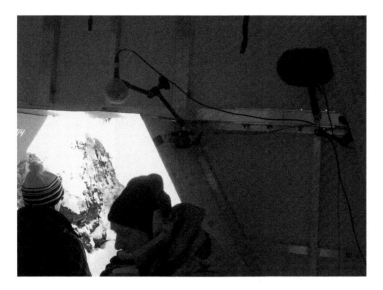

Figure 10.21 The start houses at alpine skiing offer the opportunity to create an intimate immersive experience for the viewer as the athlete prepares for competition

Figure 10.22 Fill microphones and 'playback samplers' are used on some long stretches between cameras

The finish zone uses the principles of frontal vertical soundfield enhancements which emphasizes and reinforces the front – top to bottom aspects of the mix, separating the sounds and applying elevation.

Note: The LFE channel is not used for bass management and very few sports will generate sound for the LFE. Specific microphones should be used and routed to the LFE channel and not be depended on for low frequency information in the mix.

Figure 10.23 Biathlon with aerial camera and microphone

Cross-Country and Biathlon Skiing

Cross-Country and Biathlon Skiing are long courses that are economically and physically a challenge to cover. Cross-Country and Biathlon courses are cut into the forest with a variety of twists, turns and terrain. It is common that a single camera position can cover a variety of long shots along with cameras that move with the athlete.

It is common to build at least two separate immersive sound capture configurations in the wooded area away from spectators. This provides the sonic foundation to stitch together a wide separation of camera microphones and sound supplementation from samplers. The use of samplers has been common practice by the Finnish Television Network YLE.

Cross-Country and Biathlon have stand seating at the start/finish along with fans and spectators that are in the woods along the course.

The PA gives an essential element in an immersive sound mix – height.

Case Study 10.24

Immersive Sound for Outdoor Winter Sports: Sledding and XTreme New Sports – Pipe, Cross and Acrobats

Production Philosophy for Immersive Sound

Perspective sound design is how we want to focus the attention of the viewer; for example, the motion and movement in sliding.

Unique to the sport of sledding is the close proximity of the spectators to the athlete, which provides the opportunity for a only one of its kind immersive sound design. The variety of camera angles clearly show places where the spectators are viewing from above and beside the sled on the track as well as a start and finish area with some stand seating.

The start area is a large space with seating behind the launch area. There are three primary camera views from sled level, from above and across shot with the spectators behind the sled.

There are three different immersive sound zones in sledding – start, course and finish.

At the start, the front vertical sound zone (top to bottom) reproduces driver commands and athlete banter at ear level and the driver commands echo into the above stratus either acoustically or by processing and routing. The microphones that capture the front vertical zone will also capture a distant perspective of the crowd.

These sounds are ideally oriented into the front ear level speaker and front height speakers.

Once the sled is on the course the immersive sound design moves in the soundfield as if the viewer were in the sled. The course cameras are robotic as well as operated, however most course cameras are very close to and on the track. There is a distant camera that follows a section of the course where the viewer is looking at a top view while the sled is in motion horizontally. Sound can be appropriately tied to the ear level and height speakers.

Course cameras usually have corresponding lead and trailing microphones. Elevation on lead and trailing microphones pulls the sound into the upper height channels, further contributing to the spaciousness of the soundfield.

The combination of four to six microphones at each camera position and spatial separation into the upper speakers contributes to the motion, speed and danger of the sport by sonically moving the viewer's perspective. Finally, there is a jubilant crowd at the finish area.

Outdoor sports focus the viewers' attention forward to the picture where the effects mix is contained in the front speaker's ear level and height speakers where the viewer is surrounded by atmosphere from the front and rear speakers.

There are sports, like ski jumping, that can benefit from a near and far perspective. Some effects like in-flight and landing could be mixed into a phantom middle to enhance the viewing experience.

This immersive sound design uses a variation of the principles of frontal vertical soundfield enhancements, which emphasizes and reinforces the front – top to bottom aspects of the mix and the back top to bottom aspects of the soundfield – splitting the front hemisphere and the rear hemisphere.

Note: The LFE channel is not used for bass management and very few sports will generate sound for the LFE. Specific microphones should be used and routed to the LFE channel and not be depended on for low frequency information in the mix.

Atmosphere and Landing Zones

Dynamic expansion from a subtle ambient soundfield to an expansive propagation from a properly captured ambiance and atmosphere is the basis of many outdoor immersive sound designs. Moderate silence to jubilation is an exhilarating aural effect for immersive sound.

Downhill skiing and jumping emphasizes the sonic differences in the isolation at the top of the jump, start house or on the course to the finishing and landing zones which are immersive sound zones unto their selves.

There is an artistic skill in blending the transition from "lack of sound" to an immersive soundfield.

Figure 10.24 Ambiance microphone setup for quad height layer

Start Areas: Sliding, Jumping, Cross and Aerobatics

Course: Sliding, Bobsleigh, Skeleton, Luge

Sliding events can benefit from spatial movement and trailing effects processing to emphasize the sense of speed. The Doppler shift is a natural phenomenon and there are processing algorithms that can recreate the effect.

Additionally, winter sports offer sounds that enhance the sense of height. For example, there are microphones on jumps that capture sounds that can be positioned above the listener and draw the attention to the action that appears to be above the viewer/listener.

Events where the athlete becomes airborne provide an opportunity to utilize the height channels to emphasize the motion in the soundspace above the listener.

Sports like aerials, half pipe and ski jumping have specific sounds that signal the athlete has become airborne. The microphone plans for these sports indicate that there are microphones dedicated to capturing these sounds. The lower effects mix from OBS includes Microphones A+ B + C.

Figure 10.25 Microphone cluster with luge

Clearly microphone placement can be used to enhance movement. For example, if you place a series of microphones across a panoramic space and then render the sounds across a series of speakers, the listener perceives motion as the sound source moves past and between each stationary microphone.

When the microphones are imaged between two or more real speaker positions, phantom images occur giving better sense of motion. By adding more capture points and the ability to precisely localize each source, a clearly contiguous sense of motion is achieved.

With microphone C under the control of the immersive sound mixer the sound producer can determine the ratio of bottom layer microphones A+B+C and the top layer microphones C to achieve this uplifting effect.

Aerials

This effect not only provides a connection between the layers but significantly this specific effect attracts the attention of the viewer when the athlete is airborne, enhancing the immersive experience.

Half Pipe

Half Pipe is a sport where the athlete is airborne several times during the competition and a sub-mix of the microphone operators can add dimensions in the height channels.

Immersive Sound for Sports: Live Dynamic Panning Half Pipe

The audio mixer/sound designer may have to expand their production consciousness to creatively use immersive sound. For example, the majority of the time the sound is designed to

Figure 10.26 The broadcast rights holders can request Microphones separately and separate these effects
            microphones into two separate sources

follow the camera cut of the picture, however the Half Pipe is a production where the POV of the athlete is the story. This sound design follows the athlete up the steep walls of the pipe and back down from one side to other as the athlete traverses down the Half Pipe. The up and down, side-to-side movement of the sound is the dominant sound design element and is consistent throughout the athletic competition even though the cameras will change between wide and close images. This is a design where the sound provides a consistent soundfield and dynamic motion for the picture to be cut over.

Dynamic panning is effective in the Half Pipe Production because the audio movement is the focus of attention accentuating the up and down movement of the competitor in the Half Pipe. Significantly, the point of the view of the sound is from the athlete and does not change while the athlete is in the Pipe even though the camera changes constantly.

In addition to positioning audio elements in a stationary location, there is the possibility to move sound elements through the soundfield. Dynamic spatial movement is where the audio element/object can be dynamically moved in the 3D soundfield space, in real time.

Dynamic panning and movement has tremendous creative potential and may represent a paradigm shift in the production of audio for live video production. Video production has long had digital video effects (DVE) generators where the vision mixer would move and automate graphics and pictures for a live production. Dynamic panning can be used to move sound nuggets to accentuate motion and movement and contribute to the production value of an immersive sound 4K production.

Ski jumping has specific microphones dedicated to capturing the airborne sounds of the athlete. The immersive bed is a challenge with outdoor sports. See Chapter 4, Advance audio practices.

Case Study 10.25

Immersive Sound for Free Style Skiing, Snowboard and Ski Cross

There are events such as free style skiing where there is a wider visual coverage of the athletes and course. The athlete action tends to be in the middle of the screen with the athlete navigating ramps and jumps. The athlete becomes airborne with virtually every obstacle/bump and the athlete action appearing to move between the top and bottom of the screen.

Sports sound across the top and bottom of the front soundfield tends to accentuate and complement the up and down movement of the athlete in the picture. Using the front height speaker in conjunction with the left and right speakers emphasizes the sports sounds and is beneficial to convey motion and details of the event. Free style skiing and snowboarding can benefit from spatialization of the sports sound effects into the side and height soundfield/ channels.

The audio signals is to the additional spatial signals that can be routed to channels such as the height speakers to enhance the soundfield above and to the listener. Spatialization free style skiing and snowboarding will be accomplished acoustically with microphones from the top of jump ramps.

Case Study 10.26

Immersive Sound for Wrestling

Production Philosophy for Immersive Sound

Wrestling is a sport that has been presented in full surround with sound typically dominant in the front, left and right as with most stereo productions. This stereo sound design is still good practice for the ear level soundfield in surround and in the immersive mix.

Generally wrestling has no "sports-relevant" sound in the vertical axis above the listener.

Contact microphones are placed under the floor with LFE bandpassing and splitting to the LFE. The backstage must be covered with handheld cameras and boundary microphones.

This sound design creates the complete "man-in-the-stands" soundfield where enhanced ambiance and a focused sports action sound is entertaining enough of an embellishment for this sport. Immersive sound does not need to be over the top to be effective. Sonic frequency expansion and band splitting is possible by routing microphones and combinations of microphones to separate surround or immersive group busses.

Figure 10.27 Weight Lifting has spaced pair of mono shotgun microphone in front of the weightlifter

The Host Broadcaster: The Challenges of Producing Immersive Sound for Sports from Large-scale International Events such as The Olympics and World Cup

Olympic Broadcast Services (OBS) is the Host Broadcast for the Olympic Games and is a division of the International Olympic Committee (IOC). OBS is responsible for producing the electronic coverage of the Olympic Games as well as operation of the Olympic Channel.

Part 1: The Role of the Host Broadcaster

Because of the complexity and size of hosting an Olympic Games or a World Cup the electronic television coverage of the Olympic Games is conducted by a single entity that provides a central pool coverage of audio, video, graphics and timing/scoring, essentially providing a full broadcast production to their clients also known as rights holder. The responsibility of Olympic Broadcast Services (OBS), the Host Broadcaster at the Olympics Games is to cover all athletes, every sporting event, opening and closing ceremonies and provide the rights holders with a complete production, as well as the individual elements to custom compile an individualized production suited to the world coverage of an event.

The rights holder then overlays their local identity on-top of the host broadcaster production so that it looks to their home viewers as a custom production by their rights holder. For example, in the US it looks like NBC does all the work, in Germany it looks like ARD and in Spain TVE and so on around the world; clearly the rights holders benefit from the host broadcaster core production.

Even though the host broadcaster always pushes the envelope the stereo feed is equally as important as the immersive feed because up to three-quarters of the listening audience will consume the sound in stereo. Significantly OBS provides discrete signals and mixes to the right holders.

At OBS the audio manager is responsible for the sound design as well as everything technical that impacts the audio of the entire Olympic Games. This includes cable, equipment, mounts, venue improvement (audio mixing rooms), OB Vans and the IBC – all things audio fall under the jurisdiction of the audio manager.

The Olympic Games occurs every two years, but the audio manager is working on overlapping planning cycles for several games. Finally, the audio manager spends three months on site at the game city where, during this period, the audio supervisor can find problems and variations in previous plans and hopefully accommodate all impacted parties. Efforts that should be simple like drilling holes in the floors and walls will often include venue management, equipment suppliers, sports federations and the host broadcaster audio manager to get an approval and sign-off.

Unique to the Olympics is the wide range of sports and activities conducted in a 17-day period. 60+ different disciplines, opening and closing ceremonies and, unique to the winter games, is a daily medals ceremonies and concert.

This is the first time such a wide variety of events and sports will be produced with immersive sound and with such an ambitious undertaking comes problems, solutions and opportunities. I am publishing this case study on the 2020 Tokyo Olympics with the cooperation of OBS and three months before the Tokyo Olympics.

Part 2: Philosophy: Immersive Sound Production – 2020 Tokyo Summers Olympics

From Nuno Duarte, Audio Manager and Sound Designer.

In Japan 2020 we are in the beginning of the learning curve for immersive sound; we must listen to all the sports and adjust to their own characteristics.

Figure 10.28 Nuno Duarte with Sports Federation delegates

OBS's goals are:

1. Immersive sound must be a strong new experience to the final user.
2. OBS's sound design is to cover the soundfield and NOT fill the audio soundfield with sounds. The height channels should only be used with sound elements that are part of the soundfield and not just because the sound designer now has height channels. The high channels are used mainly as the high effects, similar to the use of the LFE.
3. To accommodate all rights holders distribution systems, including current and future codecs, OBS will continue to produce discrete audio with a clear separation of the routing of the sources elements to the lower and upper lower layers. Note: there will not be any down mix upper audio source to lower audio sources.
4. The dimensional soundfield workflow must be transparent, even if the immersive effect for some sports is not the desired/expected one.
5. The immersive sound production should not increase complexity of operation for the A1/audio mixer and not increase any costs.

Unique to the Olympics is the production of a wide variety of sporting competitions, events and venues that require common sounds design and production to make a homogenous sounding broadcast production. Such an ambitious undertaking comes with problems, solutions and opportunities.

The host broadcaster is the liaison between any venue request from the rights holders such as camera, microphone and interview positions as well as booking coordination.

Host Broadcaster Sound – Overview

The host broadcast audio production and engineering team is responsible for the capture, creation and production of the audio signals and mixes. The host broadcaster has the resources to execute a comprehensive microphone plan and is capable of delivering high-quality audio stems and splits of individual effects microphones for the rights holders to produce the audio to immersive sound.

The host broadcaster is responsible for creating a complete multi-format sound mix with atmosphere, sports sounds, event- and venue-specific sound, as well as provide atmosphere and effects specific sound mixes and microphone splits to the rights holders. Separate mix groups give a rights holder the option to balance the sound of the atmosphere and sports effects according to the standards and taste of their individual countries. Additionally, quality audio stems facilitates editing and post-production of their coverage.

OBS has a history of capturing quality effects with close microphones resulting in detailed audio production even in bad acoustical environments. The close detail microphones are often mono but there is a practice of creating spaced pairs of microphones where practical and stereo microphones as often as possible.

OBS is responsible for sound to picture synchronization, which must be ensured through the entire signal chain.

Host Broadcast Mixes

Beginning at the 2020/2021 Summer Olympic Games, OBS created complete discrete mixes in immersive, surround and stereo sound. Immersive sound is defined as 9.1 or a total of 10 sound channels as designated – five channels at ear level, plus four height channels and a LFE. Note: OBS has produced a 5.1 Surround mix since 2004 and a stereo mix since 1996. OBS's mixes are basic elements of the host broadcaster and rights holders sound.

Figure 10.29 The big ball microphone

OBS is responsible for a complete audio and video package, as well as pre-booked specialty mixes and microphone splits. The complete audio mix will include event ambiance and atmosphere, event effects, OTD transition sound and any audio supplementation audio generated and added by the host broadcast mixers.

Nuno Duarte has said that:

> With Rights Holders Bookings, OBS will generate effects only mixes and atmosphere only mixes. Effects only mixes use no atmosphere microphones and no handheld camera microphones. These specialty mixes will be 4.0 because of the bus structure of the OB Vans. OBS should create a panoramic image of the venue from the camera one view.

The effects only mix is post-input fader and follows the action of the athletes and the event. This mix should be an independent stereo or surround group not in the OBS primary mix. Note These mixes should never change during the entire event.

All playback devices, loops and samplers used to augment OBS's stereo and surround sound mix should be mixed into the effect mix as a stereo or surround source.

A challenge for OBS is to create a similar sounding "soundfield" from a variety of venue acoustics.

Host Broadcast Microphone

OBS is planning to capture the present on-axis layers as well as the diffused atmosphere layers above the spectators. The general philosophy of OBS's Atmosphere capture works well for immersive sound capture. OBS has a predictable approach to atmosphere capture and often places microphones on stands at various distances in front of the crowd depending on venue construction. The ability to be able to "tune" the soundfield by adjusting the position of the microphones can optimize the atmosphere capture for various crowd sizes. Experience has proven that crowd sizes are difficult to predict for economic and political reasons.

Additionally, in some large venues, microphones are often hung from the structure of the building, ideally off walk-ways, to be able to adjust the microphones once in place to account for HVAC and PA riggings. This variety of microphones and placement facilitates the construction and collection of layers of atmosphere ranging in high detail to various amounts of diffusion creating a blend of presence and distance in a wider (sweet spot) listening space. As with all venues, OBS makes a considerable effort to get the microphones as off-axis as possible to the venue PA speakers.

A wide range of layers contribute well to an immersive soundfield and facilitate the reproduction of separate, stable and reliable atmosphere mixes that will be the foundation for all immersive sound production. Multiple layers contribute to additive/summing up-production methods as well as up-processing when rendering to an immersive sound codec. There are no plans for 3D capture from array or ambisonic microphones.

OBS audio splits, mixes and sources are stereo and surround and lack any specific height information.

OBS create a stereo and surround sound mix from the microphone sources.

The close detail microphones are often mono but there is a practice of creating spaced pairs of microphones where practical and stereo microphones as often as possible.

Stereo processing of the mono shotgun microphones to a 2 channel or 5 channel processing is a good and simple option for spatial enhancement, but all mixes are created with panning and routing and not through processing.

OBS transmission will use the 16 channels layout commonly required Dolby® Atmos™ format and this is the official format at the Olympics 2021. All host broadcast audio will be organized using 16 audio channels Dolby Atmos layout although there will be no metadata delivered by OBS as they will be transmitting immersive discrete audio.

Though all audio will be produced in an immersive audio format, broadcasters can subscribe to channel layouts from stereo all the way up to 5.1.4 with additional audio objects.

Content

The 5.1 main bed is a traditional 5.1 surround layer that contains no metadata. Even without metadata, normal panning can of course occur, though panning is likely to be static. The upper bed represents four ceiling mounted, down-pointing speakers. Also here there is no metadata, and most likely static panning.

The 5.1.4 would be containing a traditional mix of the atmosphere, for instance the ambience in a football stadium. The stereo channel(s) could, for example, contain the same ambience, but in stereo.

The objects could be as simple as just static international commentary or other static audio. As mentioned before, the transmission from the Olympics will carry no dynamic metadata, though the Dolby Atmos format supports it.

Up-Production

Music use is increasing in sports production, not only for the athlete to perform to but to excite the crowds. Greater immersion above the listener has been found to be desirable with sports production and music Up-Production using dimensional reverberation and room simulation processing black-boxes easily and convincingly converts two-dimensional audio to a three-dimensional soundfield. The dimensional channels in the soundfield create spaciousness as well as cohesion in the entire soundfield.

Music is a production element that is used as an accent and as an integral aspect to the competition such as in figure skating, floor gymnastics, basketball and beach volleyball.

Music provided by the athletes tends to be stereo and OBS will up-produce stereo to surround and immersive sound for OBS's composite immersive, surround and stereo mixes.

Music used for transitions to features or packages will be processed and up-produced from mono or stereo to surround and immersive sound for the composite immersive, surround and stereo mixes.

Ceremonies

Large-scale sporting events such as the Olympics, World Cup and Commonwealth Games typically have non-sporting pageantry events such as Ceremonies and award events like Medals Plaza. Ceremonies is a visual spectacle and is usually directed and produced over a soundtrack that is recorded and mixed in advance. Entertainment productions benefit from the dimensional soundfield of the music track. Music is the glue that ties the sound above the listener to the sound in the horizontal space and with the layered approach to the atmosphere a truly dimensional experience is achieved.

Ceremonies production is controlled by the host organizing committee who plays all voice and music tracks to OBS and simultaneously to the PA systems. The rights holders will need to have splits of all voice and vocal aspects of ceremonies production, including podium speeches, singing of anthems and live music performances. The voice aspects need to be discrete without any music pre-mixed into the voice.

Additionally, the rights holders will need all music stems after up-production to immersive and surround sound. All music will be delivered without any voice. Note: the direct music will need to be timed to the atmosphere with a slight delay to the upper soundspace.

Pyrotechnics is integral to the production and will generally be produced as a dominant 4.0 mix in the height channels.

All audio signals will be available at the Venue TOC or at the IBC. OBS has a strict policy against interference or change requests during a live broadcast by rights holders and no aspect of the host broadcast production will be compromised by request from rights holders.

Immersive Sound for Sports: Indoor Venues

An indoor venue has a distinct room tone that is contoured/shaped by the acoustic characteristics of that venue. Close microphones on the effects will hopefully minimize the room tone while good microphone placement of the atmosphere microphones will contribute to the ambiance soundfield with detail and depth.

OBS's mixes are available from every venue

1. Effects with no atmosphere
Two channel or surround

2. Atmosphere – close perspective – no effects
5.1 Surround

3. Atmosphere – distant perspective – no effects
5.1 Surround

4. PA – voice – no music
Mono

5. PA – music – no voice
Stereo

6. Microphone splits – sports specific? Any camera splits?
Mono or stereo?

7. Ambisonic microphone

8. Effects playback – non-visual
Stereo or surround

9. Visual playback with sound
Stereo or surround

All audio signals will be available at the Venue TOC or at the IBC.[6]

Immersive Sound for Sports: NHK – A Summary of Immersive Sound Development

Beside the 16 audio channel immersive audio format discussed above, NHK, the main Japanese national broadcaster, will do an additional transmission using 24 audio channels, 22.2. 22.2 is the surround sound component of the new television standard super hi-vision. It has been developed by NHK Science & Technical Research Laboratories and uses 24 speakers, including two subwoofers, arranged in three layers + LFE: Upper layer: nine channels, middle layer: ten channels, lower layer: three channels and two LFE channels. Super hi-vision will be produced from a limited number of venues and for limited distribution.

NHK is the National Broadcast of Japan and the Olympics have partnered for a number of firsts, beginning in 1964 where it was the first time Opening Ceremonies and the entire marathon was broadcast live. The 1964 games were in Japan and NHK presented almost 10 hours of live coverage each day and ever since NHK has been a leader in advance broadcasting systems sharing its expertise with the Olympic broadcasters. NHK has a long relationship with the Olympics and at promoting high broadcast standards and technology that NHK and Japanese engineers have had under development. This relationship led to NHK being the host broadcast for the 1998 games in Nagano Japan.

> 1991 – NHK became the first broadcaster to launch regular high-definition programing. Much of the high-definition footage came from the Olympics and World Cup and was acquired using NHK cameras because there were no other systems.
> 1995 – NHK demonstrated the first 8K 7680 × 4320 pixels picture with 22.2 immersive sound.
> 1996 – Mick Sawaguchi proposed six different surround sound mixing designs for drama.
> 1997 – NHK builds 5.1 surround studio in Tokyo.
> 1998 – NHK was the host broadcaster for the Nagano Winter Olympics – SD and stereo sound.
> 2002 Salt Lake – NHK broadcasts full HDTV Hi-Vision transmission of Winter Olympics with Dolby Pro Logic I – four channel audio.
> 2004 NHK selects Dolby Pro Logic II with five channels of audio because it is backwards compatible.
> 2004 AOB – Athens Olympic Broadcaster produced Opening Ceremonies, athletics, aquatics, basketball and gymnastics in surround sound and distributed Dolby Pro Logic II to NBC and NHK.
> Additionally, gymnastics – audio and video were produced by NHK.

At the 2004 Olympic Games in Athens Greece NHK played a prominent role in not only further developing and proving its own technology, but advancing global broadcast technologies and practices of the Olympics – an effort not seen since the 1936 Olympics in Berlin. In 2004 NHK dispatched to Greece Mr. Hiromi Sueishi, a senior level audio engineer, and Mr. Masato Naoe, a senior level video engineer, to implement High Definition 1920 X 1080 with 5.1 surround sound. NHK had several production and monitoring rooms in the IBC and monitored the AOB signals. In addition to supplementary crowd microphones all Ceremonies and all production music was up-produced using the TC Electronics TC6000 Unwrap algorithm from stereo to 5.1. Clearly Ceremonies music production benefitted from the up-production to 5.1.

I was the benefactor of close to two decades of experimenting and testing beginning with Mr. Mick Sarwaguchi and clearly NHK helped me learn at an accelerated pace.

> 2008 – All Summer Games Beijing produced by the host broadcaster were in HD with discrete 5.1 surround sound. Beijing Olympic broadcaster's quality control room manned by two NHK audio technicians Mr. Hinata from Athens.
>
> 2008 – 22.2 immersive audio standards set in place.
>
> NHK Sports has quietly taken a one-for-one split of large-scale microphones capture and tested many different 22.2 immersive sound compilation and production practices.
>
> 2012 – NHK produced six different 8K Channels of programming from the Summer Games in London for public viewing in theaters in Japan. During the summer games NHK moved and reset their OB Van to different venues.
>
> 2014 – During NAB NHK demonstrated "Hybridcast" system that provided interactive capabilities.

Figure 10.30 NHK Audio OB Van

2015 – Two NHK 8K OBs arrives – one for video and one for audio

2015 – July NHK broadcast baseball live from Yankee Stadium – New Audio OB

Mr. Kazutaka Noda set up 60 microphones to test capture locations.

Mr. Noda said they did not use the Big Ball because there is no good location in baseball.

NHK built – Big Ball – either a 16 or 24 microphone sphere. Must be located in the center of the venue and is susceptible to PA noise.[7]

2016 – Rio Olympics – Two side-by-side audio and video OB Vans for two separate production and technical crews. Crew A covered – opening ceremonies, swimming and basketball. Crew B covered judo, athletics, football, closing ceremonies. ENG – beach volleyball and gymnastics.

2018 – Korea – Full time presence at figure skating / speed skating, ski jumping and snowboarding – 10 camera separate video OB Vans and audio OB Van.

2020 December 1st – NHK started full time transmission/ broadcasting 4K/8K UHD content on December 1, 2018. Super Hi-Vision was scheduled to begin around the 2020 Olympics and Super Hi-Vision Cameras capable of 4K and 8K

NHK has led the way in a three-layer approach to sound design and capture. Tsuyoshi Hinata, 22.2 sound designer from the Japanese National Broadcaster, NHK, defines the use of back to front and bottom to top clarity in the aural description just as with other immersive sound formats, however 22.2 gives much more precise localization and greater perception of complete captivation.

It makes sense to add a below-ear level when you consider Blauert's description of bottom-to-top frequency banding resulting in the perception of more low frequency information at

Figure 10.31 NHK Audio OB Van view #2

ear level and below the listener and the perception of more high frequency information above the listener.

Consider the use of the frontal speakers – left low, left and left height speakers for more definition. Frontal soundfield reinforcement is described in Chapter 2. 22.2 gives the sound designer more channel options. More options are possible when placing sound elements or objects in the sonic space above and around the listener using various degrees of elevation. The 22.2 system allows for more precise degrees of elevation and separation.

There are several downsides with that 8K Super High Definition with 22.2 sound, starting with the fact that it is really not a home platform. Down converting and mixing to 11.1/5.1/stereo is complex and Fraunhofer perception research does not indicate a better experience beyond 11.1 to 22.2.

NHK and Japan has set itself up for a scalable format that has already been tested and proved.

Notes

1 Blesser, Barry, *Spaces Speak, Are You Listening? Experiencing Aural Architecture*, (Massachusetts: MIT Press, 2006).
2 Formula E Interviews with Dr. Deep Sen, Dr. Nils Peters, Dr. Martin Morrell, Qualcomm Technologies Inc., San Diego, California, USA.
3 Malone, Karl, interview, NBC Universal Sound Production – NBC Notre Dame College Football.
4 BT Sports Interview Ian Rosem, SVG Interview with Jamie Hindhaugh, Chief Operating Officer for BT Sport and BT TV.
5 Interview Simon Christian Tonemeister Fraunhofer Institute.
6 The Host Broadcaster – OBS – Olympic Broadcast Services – Interviews with Nuno Duarte Sound Manager and Designer.
7 Interviews and conversations with Mick Sawaguchi, Hiromi Sueishi, Tsuyoshi Hinata and Kazutaka Noda.

Index

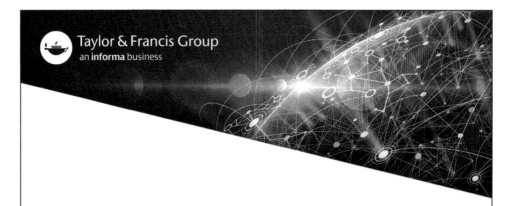